Electrical Engineering:
An Introduction

Max Kirby

NYRESEARCH
P R E S S

New York

Published by NY Research Press
118-35 Queens Blvd., Suite 400,
Forest Hills, NY 11375, USA
www.nyresearchpress.com

Electrical Engineering: An Introduction
Max Kirby

International Standard Book Number: 978-1-63238-846-9 (Hardback)

Cataloging-in-Publication Data

Electrical engineering : an introduction / Max Kirby.
     p. cm.
Includes bibliographical references and index.
ISBN 978-1-63238-846-9
1. Electrical engineering. 2. Engineering. I. Kirby, Max.
TK145 .E44 2022
621.3--dc23

Contents

Preface

The technical discipline which deals with the designing, study and application of systems or equipment that make use of electricity, electromagnetism and electronics is known as electrical engineering. Some of the sub-fields of electrical engineering are radio-frequency engineering, power engineering, instrumentation, telecommunications, signal processing and computer engineering. Radio-frequency engineering deals with the application of antenna, waveguide, transmission line and electromagnetic field principles in order to design devices which utilize or produce signals inside the radio band. The generation, transmission and distribution of electricity as well as designing related equipment are studied within power engineering. This book is a valuable compilation of topics, ranging from the basic to the most complex advancements in the field of electrical engineering. It is appropriate for students seeking detailed information in this area as well as for experts. Coherent flow of topics, student-friendly language and extensive use of examples make this book an invaluable source of knowledge.

A foreword of all chapters of the book is provided below:

Chapter 1 - The domain of engineering which focuses on the design, application and study of devices, systems and equipments which make use of electricity, electromagnetism and electronics is termed as electrical engineering. All the diverse principles of electrical engineering have been briefly introduced in this chapter.; **Chapter 2** - The energy derived from electric kinetic energy or electric potential energy is termed as electrical energy. Some of the concepts which are studied in relation to electrical energy are electrical power, electric charge and electric current. The topics elaborated in this chapter will help in gaining a better perspective about the electrical energy and its related concepts.; **Chapter 3** - Electrical circuits comprise of various components which are broadly categorized as active and passive components. The passive components are resistors, capacitors and inductors, and active components are diodes, transistors and integrated circuits. This chapter discusses in detail these components of electrical circuits.; **Chapter 4** - The circuits which comprise of one or more closed loop paths having a magnetic flux are known as magnetic circuits. There are a number of phenomena which are studied in relation to these circuits such as hysteresis and inductance. The diverse aspects of magnetic circuits and these connected phenomena have been thoroughly discussed in this chapter.; **Chapter 5** - The machines which convert electrical energy into mechanical energy or vice versa are termed as electrical machines. Some of the common electrical machines are electric generators, electric motors and transformers. The topics elaborated in this chapter will help in gaining a better perspective about these electrical machines.; **Chapter 6** - There are numerous instruments which are used to measure different quantities in electrical engineering. A few of the commonly used instruments are moving coil galvanometer, ammeter, voltmeter and multimeter. This chapter discusses in detail these instruments related to electrical engineering as well as their applications.

At the end, I would like to thank all the people associated with this book devoting their precious time and providing their valuable contributions to this book. I would also like to express my gratitude to my fellow colleagues who encouraged me throughout the process.

Max Kirby

Understanding Electrical Engineering

The domain of engineering which focuses on the design, application and study of devices, systems and equipment which make use of electricity, electromagnetism and electronics is termed as electrical engineering. All the diverse principles of electrical engineering have been briefly introduced in this chapter.

Electrical engineering (sometimes called electrical and electronics engineering) is a professional engineering discipline that deals with the development of technologies for generating and harnessing electricity for a wide range of applications. The field first became an identifiable occupation in the late nineteenth century, with the commercialization of the electric telegraph and power supply. The field now covers a range of sub-disciplines, including those that deal with power, control systems, electronics, signal processing, and telecommunications.

Electrical Engineers design power systems.

Electrical engineers may work on such things as the construction of electric power stations, the design of telecommunications systems, the wiring and lighting of buildings and transport vehicles, the design of household appliances, or the electrical control of industrial machinery. In this manner, electrical engineering plays a vital role in our modern technological society.

Tools and Work

Radar is one of many projects an electrical engineer might work on. Here, radar antennas are housed in dome-like shells called radomes.

Knowledge of physics and mathematics is fundamental to the discipline of electrical engineering, as they help provide qualitative and quantitative descriptions of how such systems will work. Today, most engineering work involves the use of computers, and it is commonplace to use computer-aided design programs when designing electrical systems.

Most electrical engineers will be familiar with basic circuit theory—that is, the interactions of elements such as resistors, capacitors, diodes, transistors, and inductors in a circuit. In addition, engineers rely on theories that are more specific to the type of work they do. For example, quantum mechanics and solid state physics might be relevant to an engineer working in microelectronics, but they are largely irrelevant to engineers working with macroscopic electrical systems. Even circuit theory may not be relevant to a person designing telecommunications systems that use commercial, off-the-shelf components. Perhaps the most important technical skills for electrical engineers are reflected in university programs, which emphasize strong numerical skills, computer literacy, and the ability to understand the technical language and concepts related to electrical engineering.

For most engineers, technical work accounts for only a fraction of their job. Much time is spent on tasks such as discussing proposals with clients, preparing budgets, and determining project schedules. Many senior engineers manage a team of technicians or other engineers, and for this reason project management skills are important. In addition, most engineering projects involve producing some form of documentation, requiring strong written communication skills.

The workplaces of electrical engineers are just as varied as the types of work they do. They may be in a pristine lab environment in a fabrication plant, the offices of a consulting firm, or on-site at a mine. They may find themselves supervising a wide range of individuals, including scientists, electricians, computer programmers, and other engineers.

Subdisciplines

Electrical engineering has many branches or subdisciplines, the most popular of which are listed below. Although some electrical engineers focus exclusively on one or other subdiscipline, many deal with several branches. Some fields, such as electronics engineering and computer engineering, are considered separate disciplines in their own right.

Power Engineering

Power engineering deals with the generation, transmission, and distribution of electricity. It includes the design of a range of devices, such as transformers, electric generators, electric motors, and power electronics. In many parts of the world, governments maintain electrical networks called power grids, which connect electric generators with users. By purchasing electrical energy from the grid, consumers can avoid the high cost of generating their own. Power engineers may work on the design and maintenance of the power grid as well as the power systems connected to it. Such *on-grid* power systems may supply the grid with additional power, draw power from the grid, or do both. Power engineers may also work on *off-grid* power systems, which are not connected to the grid and may, in some cases, be preferable to on-grid systems.

Control Engineering

Launching of the NASA space shuttle Columbia.

Control engineering focuses on the modeling of a diverse range of dynamic systems and the design of controllers that will cause these systems to behave as desired. To implement such controllers, electrical engineers may use electrical circuits, digital signal processors, and microcontrollers. Control engineering has a wide range of applications, from the flight and propulsion systems of spacecraft and airliners to the cruise control in modern automobiles. It also plays an important role in industrial automation.

When designing control systems, control engineers often utilize feedback. For example, in an automobile with cruise control, the vehicle's speed is continuously monitored and fed back to the system, which adjusts the motor's speed accordingly. In cases of regular feedback, control theory can be used to determine how the system responds to such feedback.

Electronics Engineering

Electronics engineering involves the design and testing of electronic circuits that use the properties of components such as resistors, capacitors, inductors, diodes, and transistors to obtain particular functions. The tuned circuit, which allows the radio user to filter out all but a single station, is one example of such a circuit.

A transducer circuit.

Prior to World War II, the subject was commonly known as radio engineering and basically was restricted to radar and some aspects of communications, such as commercial radio and early television. In the post-war years, as consumer devices began to be developed, the field grew to include modern television, audio systems, computers, and microprocessors. In the mid to late 1950s, the term radio engineering gradually gave way to the name electronics engineering.

Before invention of the integrated circuit in 1959, electronic circuits were constructed from discrete components that could be manipulated by people. These discrete circuits, still common in some applications, consumed much space and power and were limited in speed. By contrast, integrated circuits packed a large number—often millions—of tiny electrical components, mainly transistors, into a small chip around the size of a coin. This innovation allowed for the powerful computers and other electronic devices we have today.

Microelectronics

Microelectronics engineering deals with the design of extremely small (microscopic) electronic components for use in an integrated circuit, or occasionally for use on their own as general electronic components. The most common microelectronic components are semiconductor transistors, but all main electronic components (resistors, capacitors, inductors) can be made on the microscopic level.

Most components are designed by determining processes for mixing silicon with other chemical elements to create the desired electromagnetic effect. For this reason, microelectronics involves a significant amount of quantum mechanics and chemistry.

Signal Processing

Signal processing deals with the analysis and manipulation of signals, which may be analog or digital. An analog signal varies continuously according to the information carried, and a digital signal varies according to a series of discrete values that represent the information. Signal processing of analog signals may involve the amplification and filtering of audio signals for audio equipment, or the modulation and demodulation of signals for telecommunications. In the case of digital signals, signal processing may involve the compression, error detection, and error correction of digitally sampled signals.

Telecommunications

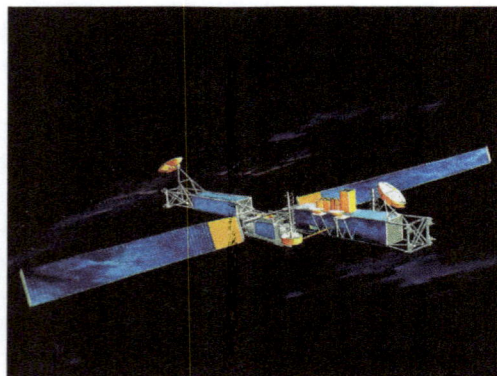

Milstar communications satellite.

Telecommunications engineering focuses on the transmission of information across a channel such as a coax cable, optical fiber, or free space. Transmissions across free space require information to be encoded in a carrier wave, to shift the information to a carrier frequency suitable for transmission; this is known as modulation. Popular analog modulation techniques include amplitude modulation (AM) and frequency modulation (FM). The choice of modulation affects the cost and performance of a system, and the engineer must carefully balance these two factors.

Once a system's transmission characteristics are determined, telecommunications engineers design the transmitters and receivers needed for such systems. These two are sometimes combined to form a two-way communication device known as a transceiver. A key consideration in the design of transmitters is their power consumption, which is closely related to their signal strength. If a transmitter's signal strength is insufficient, the signal's information will be corrupted by noise.

Instrumentation Engineering

Radar gun.

Instrumentation engineering deals with the design of devices to measure physical quantities, such as pressure, flow, and temperature. The design of such instrumentation requires a good understanding of physics, often extending beyond electromagnetic theory. For example, radar guns use the Doppler Effect to measure the speed of oncoming vehicles. Similarly, thermocouples use the Peltier-Seebeck effect to measure the temperature difference between two points.

Often, the devices are not used by themselves but may act as sensors in larger electrical systems. For example, a thermocouple may be used to help ensure that the temperature of a furnace remains constant. From this perspective, instrumentation engineering is often viewed as the counterpart of control engineering.

Computer Engineering

Personal digital assistant.

Computer engineering deals with the design of computers and computer systems. It may involve the design of new hardware, the design of personal digital assistants (PDAs), or the use of computers to control an industrial plant. Computer engineers may also work on a system's software, although the design of complex software systems is often the domain of software engineering, which is usually considered a separate discipline. Desktop computers represent a tiny fraction of the devices a computer engineer might work on, as computer-like architectures are now found in a range of devices, including video game consoles and DVD players.

Related Disciplines

Mechatronics is an engineering discipline that deals with the convergence of electrical and mechanical systems. Such combined systems are known as electromechanical systems and are widely used. Examples include automated manufacturing systems; heating, ventilation, and air-conditioning systems (HVAC); and various subsystems of aircraft and automobiles.

The term mechatronics is typically used to refer to macroscopic systems, but futurists have predicted the emergence of very small electromechanical devices. Already such small devices—known as micro-electromechanical systems (MEMS)—are used in automobiles to tell airbags when to deploy, digital projectors to create sharper images, and inkjet printers to create nozzles for high-definition printing.

Biomedical engineering is another related discipline. It is concerned with the design of medical equipment, including (a) fixed equipment, such as ventilators, MRI scanners, and electrocardiograph monitors, and (b) mobile equipment, such as cochlear implants, artificial pacemakers, and artificial hearts.

Electrical Energy

The energy derived from electric kinetic energy or electric potential energy is termed as electrical energy. Some of the concepts which are studied in relation to electrical energy are electrical power, electric charge and electric current. The topics elaborated in this chapter will help in gaining a better perspective about the electrical energy and its related concepts.

The energy which is caused by the movement of the electrons from one place to another such type of energy is called electrical energy. In other words, electrical energy is the work done by the moving streams of the electrons or charges. Electrical energy is the form of kinetic energy because it produces by the movement of the electrical charges. The faster the movement of charges the more the energy they carry.

Considered a circuit shown in the figure below. When a potential difference P is applied across the circuit a current (I amperes) flow through it for a particular period t seconds. The voltage applied across the circuit is equal to the ratio of the work done by the electrical charge to the number of electrical charges present in the circuit. It is expressed by the formula shown below.

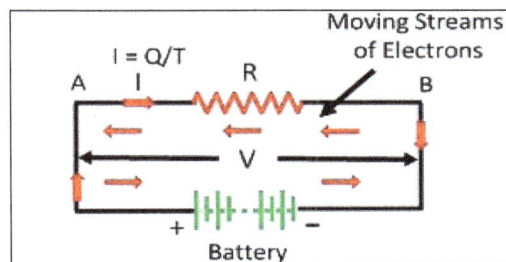

$$V = \frac{Work\ Done}{Q}$$

Therefore work done or electrical energy expanded:

$$Work\ done = VQ$$

Since,

$$I = \frac{Q}{t}$$

$$Wor\ done\ = VIt$$
$$Work\ done = I^2Rt$$
$$Work\ done = \frac{V^2}{R}t$$

Unit of Electrical Energy

The basic unit of the electrical energy is the joule (or watt-second). If the voltage is equal to the one volt, the current is equal to the one ampere and the time is equal to the one second then the electrical energy is equal to the one joule.

Hence the energy expended in an electrical circuit is said to be one joule (or watt second) if one-ampere current flows through the circuit for one second when the potential difference of one volt is applied across it.

The commercial or practical unit of energy is the kilowatt-hour (kWh) which is also known as the Board of Trade (B.O.T) unit.

$$lkWh = 1000 \times 60 \times 60 \; watt - \text{seconds}$$

$$1kWh = 36 \times 10^5 Ws \; or \; joules$$

Usually, one kWh is called one unit.

Electrical Power

The rate at which the work is being done in an electrical circuit is called an electric power. In other words, the electric power is defined as the rate of the transferred of energy. The electric power is produced by the generator and can also be supplied by the electrical batteries. It gives a low entropy form of energy which is carried over long distance and also it is converted into various other forms of energy like motion, heat energy, etc.

The electric power is divided into two types, i.e., the AC power and the DC power. The classification of the electric power depends on the nature of the current. The electric power is sold regarding joule which is the product of the power in kilowatts and the running time of the machinery in hours. The utility of power is measured by the electric meter which records the total energy consumed by the powered devices. The electric power is given by the equation shown below.

$$Electrical \; Power = \frac{Work \; done \; in \; an \; electrical \; current}{time}$$

$$P = \frac{VIt}{t} = VI = IR^2 = \frac{V^2}{R}$$

Where, V is the voltage in volts, I is the current in amperes, R is the resistance offered by the powered devices, T is the time in seconds and the P is the power measured in watts.

Unit of Electric Power

The unit of electrical power is Watt.

If,

$$V = 1 volts \text{ and } I = 1 \text{ ampere}$$
$$P = 1 \text{ watt}$$

Thus, the power consumed in an electrical circuit is said to one watt if one ampere current flows through the circuit when a potential difference of 1 volt is applied across it. The bigger unit of electrical power is the kilowatt (kW), it is usually used in the power system

$$1kW = 1000W$$

Types of an Electric Power

The electrical power is mainly classified into two types. They are the DC power and the AC power.

DC Power

The DC power is defined as the product of the voltage and current. It is produced by the fuel cell, battery and generator.

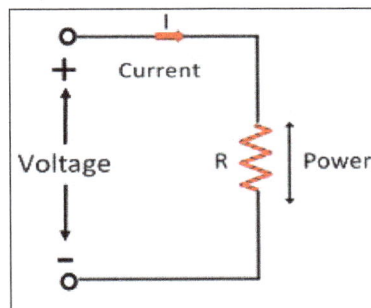

Electric Power.

$$P = V \times I$$

Where,

P – Power in watt.

V – Voltage in volts.

I – current in amps.

AC Power

The AC power is mainly classified into three types. They are the apparent power, active power and real power.

1. Apparent Power: The apparent power is the useless power or idle power. It is represented by the symbol S, and their SI unit is volt-amp.

$$S = V_{rms} I_{rms}$$

Where,

S – Apparent power.

V_{rms} – RMS voltage = $V_{peak}\sqrt{2}$ in volt.

I_{rms} – RMS current = $I_{peak}\sqrt{2}$ in the amp.

2. Active Power: The active power (P) is the real power which is dissipated in the circuit resistance.

$$P = V_{max}\, I_{max}\, Cos\phi$$

Where,

P – the real power in watts.

V_{rms} – RMS voltage = $V_{peak}\sqrt{2}$ in volts.

I_{rms} – RMS current = $I_{peak}\sqrt{2}$ in the amp.

Φ – impedance phase angle between voltage and current.

3. Reactive Power: The power developed in the circuit reactance is called reactive power (Q). It is measured in volt-ampere reactive.

$$Q = V_{rms}I_{rms}Sin\phi$$

Where,

Q – The reactive power in watts.

V_{rms} – RMS voltage = $V_{peak}\sqrt{2}$ in volt.

I_{rms} – RMS current = $I_{peak}\sqrt{2}$ in the amp.

Φ – Impedance phase angle between voltage and current.

The relation between the apparent, active and reactive power is shown below.

$$S^2 = Q^2 + P^2$$

The ratio of the real to the apparent power is called power factor, and their value lies between 0 and 1.

Power Ratings

All electronic components transfer energy from one type to another. Some energy transfers are desired: LEDs emitting light, motors spinning, batteries charging. Other energy transfers are undesirable, but also unavoidable. These unwanted energy transfers are power losses, which usually show up in the form of heat. Too much power loss - too much heat on a component - can become very undesirable.

Even when energy transfers are the main goal of a component, there'll still be losses to other forms of energy. LEDs and motors, for example, will still produce heat as a byproduct of their other energy transfers.

Most components have a rating for maximum power they can dissipate, and it's important to keep them operating under that value.

Resistor Power Ratings

Resistors are some of the more notorious culprits of power loss. When you drop some voltage across a resistor, you're also going to induce current flow across it. More voltage, means more current, means more power.

If 9V were dropped across a 10Ω resistor, that resistor would dissipate 8.1W. 8.1 is a lot of watts for most resistors. Most resistors are rated for anywhere from ⅛W (0.125W) to ½W (0.5W). If you drop 8W across a standard ½W resistor, ready a fire extinguisher.

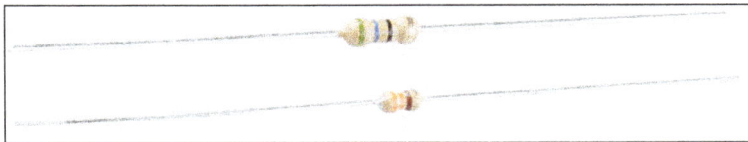

If you've seen resistors before, you've probably seen these. Top is a ½W resistor
and below that a ¼W.

These aren't built to dissipate very much power.

There are resistors built to handle large power drops. These are specifically called out as power resistors.

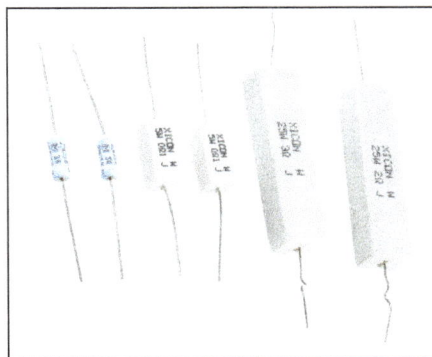

These large resistors are built to dissipate lots of power. From left to right: two 3W 22kΩ resistors,
two 5W 0.1Ω resistors, and 25W 3Ω and 2Ω resistors.

If you ever find yourself picking out a resistor value. Keeps its power rating in mind as well. And, unless your goal is to heat something up (heating elements are basically really high-power resistors), try to minimize power loss in a resistor.

Resistor power ratings can come into play when you're trying to decide on a value for an LED current-limiting resistor. Say, for example, you want to light up a 10mm super-bright red LED at maximum brightness, using a 9V battery.

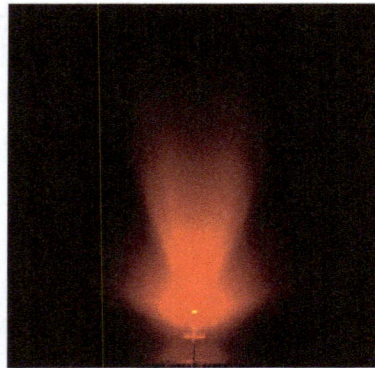

That LED has a maximum forward current of 80mA, and a forward voltage of about 2.2V. So to deliver 80mA to the LED, you'd need an 85Ω resistor to do so.

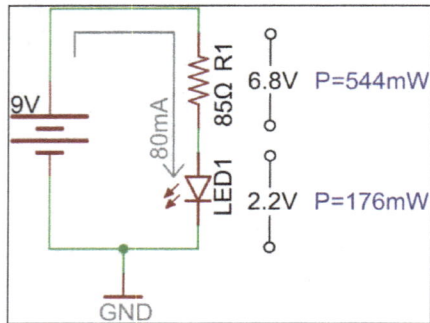

6.8V dropped on the resistor, and 80mA running through it means 0.544W (6.8V*0.08A) of power lost on it. A half-watt resistor isn't going to like that very much! It probably won't melt, but it'll get hot. Play it safe and move up to a 1W resistor (or save power and use a dedicated LED driver).

Resistors certainly aren't the only components where maximum power ratings must be considered. Any component with a resistive property to it is going to produce thermal power losses. Working with components that are commonly subjected to high power -- voltage regulators, diodes, amplifiers, and motor drivers, for example -- means paying extra special attention to power loss and thermal stress.

Electric Charge

Benjamin Franklin was the first American scientist who proved that there are two types of electric charges: positive charge and negative charge.

Electric charge is the property of sub-atomic particles particularly includes electrons and protons. Electrons have negative charge, protons have positive charge and neutrons do not have any charge. The charge of electrons and protons is measured in coulombs, represented by C. Electron has a charge of -1.602×10^{-19} Coulombs (C) and proton has a charge of $+1.602 \times 10^{-19}$ Coulombs (C). The charge of an electron is equal to the charge of a proton. However, electron has a negative value of charge and proton has a positive value of charge.

Generally, the number of electrons and protons in the atom are equal in number. Due to the opposite charges of electrons and protons the charges get cancel each other and the atom remains neutral.

However, if the atom has unequal number of electrons and protons, then the atom is said to be a charged atom. If the atom has more number of electrons (negative charges) than protons (positive charges), then it is said to be negatively charged. Similarly, if the atom has more number of protons than electrons, then it is said to be positively charged.

Properties of Electric Charge

The various properties of electric charge include:

- Additivity of charges

- Charge is conserved

- Quantization of charge

Additivity of Charges

If a system contains two point charges q_1 and q_2, then the total charge of the system is obtained by simply adding q_1 and q_2, i.e., charges add up like real numbers.

If a system contains n number of charges $q_1, q_2, q_3, q_4, ------, q_n$, then the total charge of the system is $q_1 + q_2 + q_3 + q_4 + --------- + q_n$.

Charge is a scalar quantity; it has magnitude but no direction, similar to mass. However, there is one difference between charge and mass. Mass of a body is always positive whereas charge can be either positive or negative.

Let us take for example, the system containing four charges $q_1 = +2C$, $q_2 = +3C$, $q_3 = -3C$, $q_4 = +4C$, then the total charge of the system is:

$$q = q_1 + q_2 + q_3 + q_4$$

$$= (+2) + (+3) + (-3) + (+4)$$

$$= +6C$$

Therefore, the total charge of the system is +6C and it is positively charged.

Charge is Conserved

The law of conservation of charge states that charge cannot be created or destroyed. However, a charge can be transferred from one object to other.

Let us consider two objects, object A and object B. Object A has equal number of electrons and protons. So, it is electrically neutral. Similarly, object B has equal number of electrons and protons. So, it is also electrically neutral.

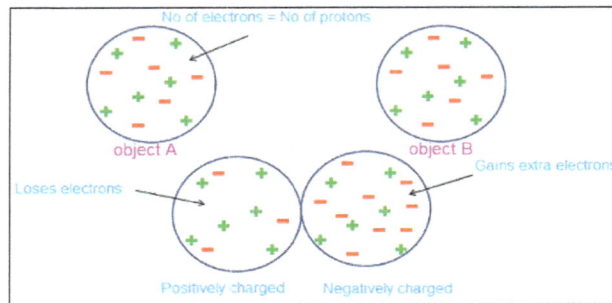

When object A and object B are rubbed with each other, negative charges from object A can be transferred to object B. Hence, object B has more number of electrons than protons due to gaining of extra electrons. Similarly, object A has lesser number of electrons than protons due to losing of some electrons.

Therefore, object A becomes positively charged and object B becomes negatively charged. However, the total charge of an isolated system remains constant.

Quantization of Charge

The charge of any object is equal to integer multiples of the elementary charge. This is known as quantization of charge. It is given by:

$$q = ne \text{ (or) } n(-e)$$

Where,

q = electric charge of any object or body.

n = any integer positive or negative.

-e = elementary charge = charge carried by single electron.

e = elementary charge = charge carried by single proton.

The charge on an electron is written as $-e$ and charge on a proton is written as $+e$. The quantization of charge was first suggested by the experimental laws of electrolysis discovered by Faraday. It was experimentally proved by Millikan. The total charge on a object is equal to the algebraic sum of individual charges present within the object.

If an object contains n_1 electrons and n_2 protons, then the total charge on the object is $n_1 \times (-e) + n_2 \times e$. For example, if the object contains 150 electrons and 200 protons, then the total charge on the object is $-150e + 200e = 50e$. Hence, the object is positively charged. The object charge can be exactly $0e$ or $1e$, $2e$ - - - or $-1e$, $-2e$ - - - but not $1/2$, $1/4$ etc.

Methods of Charging

The process of supplying the electric charge (electrons) to an object or losing the electric charge (electrons) from an object is called charging. An uncharged object can be charged in different ways.

- Charging by friction

- Charging by conduction

- Charging by induction

Charging by Friction

When an object is rubbed over another object, the electrons get transferred from one object to another. This transfer of electrons takes place due to friction between the two objects. The object that transfers electrons loses negative charge (electrons) and the object that accepts electrons gains negative charge (electrons).

Hence, the object that gains extra electrons becomes negatively charged and the object that loses electrons becomes positively charged. Thus, the two objects get charged by friction. The charge obtained on the two objects is called friction charge. This method of charging an object is called electrification by friction.

Charging by Conduction

The process of charging the uncharged object by bringing it in contact with another charged object is called charging by conduction.

A charged object has unequal number of negative (electrons) and positive charges (protons). Hence, when a charged object is brought in contact with the uncharged conductor, the electrons get transferred from charged object to the conductor.

Consider an uncharged metal rod A kept on an insulating stand and a negatively charged conductor B as shown in below figure.

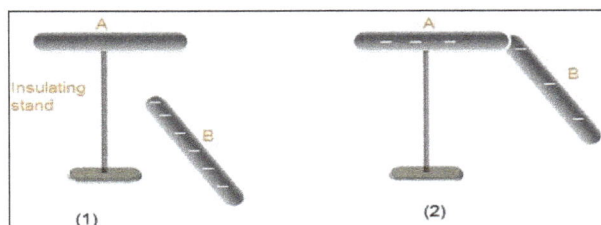

If we touch the uncharged conductor A with the negatively charged conductor B, transfer of electrons from charged conductor to uncharged conductor takes place. Hence, uncharged conductor gains extra electrons and charged conductor loses electrons. Thus, uncharged conductor A becomes negatively charged by gaining of extra electrons.

Similarly, uncharged conductor becomes positively charged if it is brought in contact with positively charged conductor.

Charging by Induction

The process of charging the uncharged object by bringing another charged object near to it, but not touching it, is called charging by induction.

Consider an uncharged metal sphere and negatively charged plastic rod as shown in below figure. If we bring the negatively charged plastic rod near to uncharged sphere as shown in below figure, charge separation occurs.

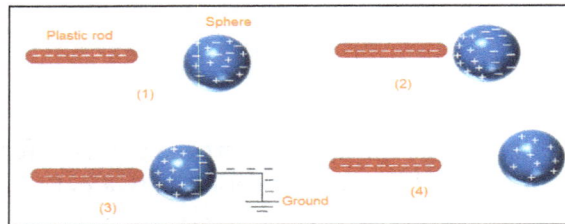

The positive charges in the sphere get attracted towards the plastic rod and move to one end of the sphere that is closer to the plastic rod. Similarly, negative charges get repelled from the plastic rod and move to another end of the sphere that is farther away from the plastic rod. Thus, the charges in the sphere rearrange themselves in a way that all the positive charges are nearer to the plastic rod and all the negative charges are farther away from it.

If this sphere is connected to a ground through the wire as shown in figure, free electrons of the sphere at farther end flow to the ground. Thus, the sphere becomes positively charged by induction. If the plastic rod is removed as shown in fig all the positive charges spread uniformly in the sphere.

Electricity

Electricity is the phenomenon associated with stationary or moving electric charges. Electric charge is a fundamental property of matter and is borne by elementary particles. In electricity the particle involved is the electron, which carries a charge designated, by convention, as negative. Thus, the various manifestations of electricity are the result of the accumulation or motion of numbers of electrons.

Electrostatics

Electrostatics is the study of electromagnetic phenomena that occur when there are no moving charges—i.e., after a static equilibrium has been established. Charges reach their equilibrium positions rapidly because the electric force is extremely strong. The mathematical methods of electrostatics make it possible to calculate the distributions of the electric field and of the electric potential from a known configuration of charges, conductors, and insulators. Conversely, given a set of conductors with known potentials, it is possible to calculate electric fields in regions between the conductors and to determine the charge distribution on the surface of the conductors. The electric energy of a set of charges at rest can be viewed from the standpoint of the work required to assemble the charges; alternatively, the energy also can be considered to reside in the electric field produced by this assembly of charges. Finally, energy can be stored in a capacitor; the energy required to charge such a device is stored in it as electrostatic energy of the electric field.

Coulomb's Law

Static electricity is a familiar electric phenomenon in which charged particles are transferred from one body to another. For example, if two objects are rubbed together, especially if the objects are insulators and the surrounding air is dry, the objects acquire equal and opposite charges and an attractive force develops between them. The object that loses electrons becomes positively charged, and the other becomes negatively charged. The force is simply the attraction between charges of opposite sign. The properties of this force are incorporated in the mathematical relationship known as Coulomb's law. The electric force on a charge Q_1 under these conditions, due to a charge Q_2 at a distance r, is given by Coulomb's law,

$$F = k\frac{Q_1 Q_2}{r^2}\hat{r}.$$

The bold characters in the equation indicate the vector nature of the force, and the unit vector $\hat{r}$ is a vector that has a size of one and that point from charge Q_2 to charge Q_1. The proportionality constant k equals $10^{-7}c^2$, where c is the speed of light in a vacuum; k has the numerical value of 8.99×10^9 newtons-square metre per coulomb squared (Nm^2/C^2). Figure shows the force on Q_1 due to Q_2. A numerical example will help to illustrate this force. Both Q_1 and Q_2 are chosen arbitrarily to be positive charges, each with a magnitude of 10^{-6} coulomb. The charge Q_1 is located at coordinates x, y, z with values of 0.03, 0, 0, respectively, while Q_2 has coordinates 0, 0.04, 0. All coordinates are given in metres. Thus, the distance between Q_1 and Q_2 is 0.05 metre.

The magnitude of the force F on charge Q_1 as calculated using equation $F = k\frac{Q_1 Q_2}{r^2}\hat{r}$ is 3.6 newtons; its direction is shown in figure. The force on Q_2 due to Q_1 is −F, which also has a magnitude of 3.6 newtons; its direction, however, is opposite to that of F. The force F can be expressed in terms of its components along the x and y axes, since the force vector lies in the xy plane. This is done with elementary trigonometry from the geometry of Figure, and the results are shown in figure. Thus,

$$F = 2.16\hat{x} - 2.88\hat{y},$$

in newtons. Coulomb's law describes mathematically the properties of the electric force between charges at rest. If the charges have opposite signs, the force would be attractive; the attraction would be indicated in equation $F = k\frac{Q_1 Q_2}{r^2}\hat{r}$ by the negative coefficient of the unit vector $\hat{r}$. Thus, the electric force on Q_1 would have a direction opposite to the unit vector $\hat{r}$ and would point from Q_1 to Q_2. In Cartesian coordinates, this would result in a change of the signs of both the x and y components of the force in equation $F = 2.16\hat{x} - 2.88\hat{y}$.

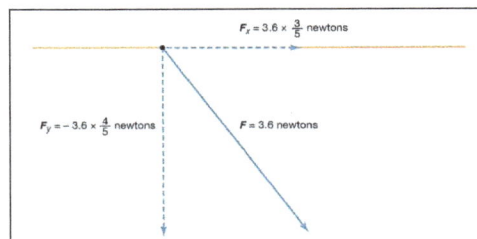

The x and y components of the force F.

How can this electric force on Q_1 be understood? Fundamentally, the force is due to the presence of an electric field at the position of Q_1. The field is caused by the second charge Q_2 and has a magnitude proportional to the size of Q_2. In interacting with this field, the first charge some distance away is either attracted to or repelled from the second charge, depending on the sign of the first charge.

Calculating the Value of an Electric Field

In the example, the charge Q_1 is in the electric field produced by the charge Q_2. This field has the value,

$$E = k\frac{Q_2}{r^2}\hat{r}$$

in newtons per coulomb (N/C). (Electric field can also be expressed in volts per metre [V/m], which is the equivalent of newtons per coulomb.) The electric force on Q_1 is given by,

$$F = 2.16\hat{x} - 2.88\hat{y}$$

in newtons. This equation can be used to define the electric field of a point charge. The electric field E produced by charge Q_2 is a vector. The magnitude of the field varies inversely as the square of the distance from Q_2; its direction is away from Q_2 when Q_2 is a positive charge and toward Q_2 when Q_2 is a negative charge. The field produced by Q_2 at the position of Q_1 is:

$$F = Q_1 E$$

$$E = 2.16 \times 10^6 \hat{x} - 2.88 \times 10^6 \hat{y}$$

in newtons per coulomb.

When there are several charges present, the force on a given charge Q_1 may be simply calculated as the sum of the individual forces due to the other charges Q_2, Q_3, etc., until all the charges are included. This sum requires that special attention be given to the direction of the individual forces since forces are vectors. The force on Q_1 can be obtained with the same amount of effort by first calculating the electric field at the position of Q_1 due to Q_2, Q_3, etc. To illustrate this, a third charge is added to the example above. There are now three charges, $Q_1 = +10^{-6}$ C, $Q_2 = +10^{-6}$ C, and $Q_3 = -10^{-6}$ C. The locations of the charges, using Cartesian coordinates [x, y, z] are, respectively, [0.03, 0, 0], [0, 0.04, 0], and [−0.02, 0, 0] metre, as shown in figure. The goal is to find the force on Q_1. From the sign of the charges, it can be seen that Q_1 is repelled by Q_2 and attracted by Q_3. It is also clear that these two forces act along different directions. The electric field at the position of Q_1 due to charge Q_2 is, just as in the example above,

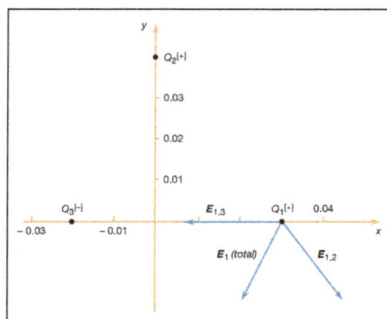

Electric field at the location of Q_1.

$$E_{1,2} = 2.16 \times 10^6 \hat{x} - 2.88 \times 10^6 \hat{y}$$

in newtons per coulomb. The electric field at the location of Q_1 due to charge Q_3 is:

$$E_{1,3} = -3.6 \times 10^8 \hat{x}$$

in newtons per coulomb. Thus, the total electric field at position 1 (i.e., at [0.03, 0, 0]) is the sum of these two fields $E_{1,2} + E_{1,3}$ and is given by,

$$E_1\left(total\right) = -1.44 \times 10^6 \hat{x} - 2.88 \times 10^6 \hat{y}$$

The fields $E_{1,2}$ and $E_{1,3}$, as well as their sum, the total electric field at the location of Q_1, E_1 (total), are shown in figure. The total force on Q_1 is then obtained from equation $F = Q_1 E$ by multiplying the electric field E_1 (total) by Q_1. In Cartesian coordinates, this force, expressed in newtons, is given by its components along the x and y axes by,

$$F_{1(total)} = -1.44 \hat{x} - 2.88 \hat{y}.$$

The resulting force on Q_1 is in the direction of the total electric field at Q_1, shown in figure. The magnitude of the force, which is obtained as the square root of the sum of the squares of the components of the force given in the above equation, equals 3.22 newtons.

Superposition Principle

This calculation demonstrates an important property of the electromagnetic field known as the superposition principle. According to this principle, a field arising from a number of sources is determined by adding the individual fields from each source. The principle in which an electric field arising from several sources is determined by the superposition of the fields from each of the sources. In this case, the electric field at the location of Q_1 is the sum of the fields due to Q_2 and Q_3. Studies of electric fields over an extremely wide range of magnitudes have established the validity of the superposition principle.

The vector nature of an electric field produced by a set of charges introduces a significant complexity. Specifying the field at each point in space requires giving both the magnitude and the direction at each location. In the Cartesian coordinate system, this necessitates knowing the magnitude of the x, y, and z components of the electric field at each point in space. It would be much simpler if the value of the electric field vector at any point in space could be derived from a scalar function with magnitude and sign.

Electric Potential

The electric potential is just such a scalar function. Electric potential is related to the work done by an external force when it transports a charge slowly from one position to another in an environment containing other charges at rest. The difference between the potential at point A and the potential at point B is defined by the equation,

$$V_A - V_B = \frac{\text{work to move charge q from B to A}}{q}.$$

As noted above, electric potential is measured in volts. Since work is measured in joules in the Système Internationale d'Unités (SI), one volt is equivalent to one joule per coulomb. The charge q is taken as a small test charge; it is assumed that the test charge does not disturb the distribution of the remaining charges during its transport from point B to point A.

To illustrate the work in equation $V_A - V_B = \dfrac{\text{work to move charge q from B to A}}{q}$, figure shows a positive charge +Q. Consider the work involved in moving a second charge q from B to A. Along path 1, work is done to offset the electric repulsion between the two charges. If path 2 is chosen instead, no work is done in moving q from B to C, since the motion is perpendicular to the electric force; moving q from C to D, the work is, by symmetry, identical as from B to A, and no work is required from D to A. Thus, the total work done in moving q from B to A is the same for either path. It can be shown easily that the same is true for any path going from B to A. When the initial and final positions of the charge q are located on a sphere centred on the location of the +Q charge, no work is done; the electric potential at the initial position has the same value as at the final position. The sphere in this

example is called an equipotential surface. When equation $V_A - V_B = \dfrac{\text{work to move charge q from B to A}}{q}$,

which defines the potential difference between two points, is combined with Coulomb's law, it yields the following expression for the potential difference $V_A - V_B$ between points A and B:

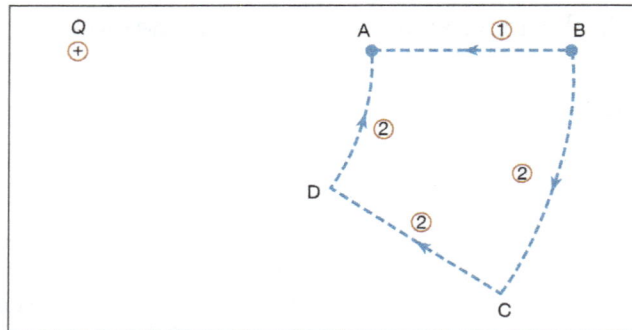

Positive charge +Q and two paths in moving a second charge, q, from B to A.

$$V_A - V_B = k\frac{Q}{r_a} - k\frac{Q}{r_b}.$$

Where, r_a and r_b are the distances of points A and B from Q. Choosing B far away from the charge Q and arbitrarily setting the electric potential to be zero far from the charge results in a simple equation for the potential at A:

$$V_A = k\frac{Q}{r_a}.$$

The contribution of a charge to the electric potential at some point in space is thus a scalar quantity directly proportional to the magnitude of the charge and inversely proportional to the distance between the point and the charge. For more than one charge, one simply adds the contributions of the various charges. The result is a topological map that gives a value of the electric potential for every point in space.

Figure provides three-dimensional views illustrating the effect of the positive charge +Q located at the origin on either a second positive charge q or on a negative charge –q; the potential energy "landscape" is illustrated in each case. The potential energy of a charge q is the product qV of the charge and of the electric potential at the position of the charge. In figure, the positive charge q would have to be pushed by some external agent in order to get close to the location of +Q because, as q approaches, it is subjected to an increasingly repulsive electric force. For the negative charge –q, the potential energy in figure shows, instead of a steep hill, a deep funnel. The electric potential due to +Q is still positive, but the potential energy is negative, and the negative charge –q, in a manner quite analogous to a particle under the influence of gravity, is attracted toward the origin where charge +Q is located.

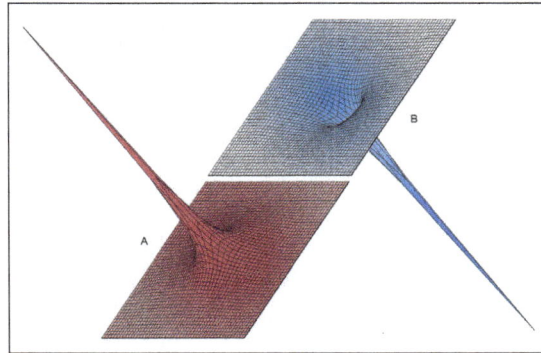

Potential energy landscape. (A) Potential energy of a positive charge near a second positive charge. (B) Potential energy of a negative charge near a positive charge.

The electric field is related to the variation of the electric potential in space. The potential provides a convenient tool for solving a wide variety of problems in electrostatics. In a region of space where the potential varies, a charge is subjected to an electric force. For a positive charge the direction of this force is opposite the gradient of the potential—that is to say, in the direction in which the potential decreases the most rapidly. A negative charge would be subjected to a force in the direction of the most rapid increase of the potential. In both instances, the magnitude of the force is proportional to the rate of change of the potential in the indicated directions. If the potential in a region of space is constant, there is no force on either positive or negative charge. In a 12-volt car battery, positive charges would tend to move away from the positive terminal and toward the negative terminal, while negative charges would tend to move in the opposite direction—i.e., from the negative to the positive terminal. The latter occurs when a copper wire, in which there are electrons that are free to move, is connected between the two terminals of the battery.

Deriving Electric Field from Potential

The electric field has already been described in terms of the force on a charge. If the electric potential is known at every point in a region of space, the electric field can be derived from the potential. In vector calculus notation, the electric field is given by the negative of the gradient of the electric potential, E = –grad V. This expression specifies how the electric field is calculated at a given point. Since the field is a vector, it has both a direction and magnitude. The direction is that in which the potential decreases most rapidly, moving away from the point. The magnitude of the field is the change in potential across a small distance in the indicated direction divided by that distance.

To become more familiar with the electric potential, a numerically determined solution is presented for a two-dimensional configuration of electrodes. A long, circular conducting rod is maintained at an electric potential of −20 volts. Next to the rod, a long L-shaped bracket, also made of conducting material, is maintained at a potential of +20 volts. Both the rod and bracket are placed inside a long, hollow metal tube with a square cross section; this enclosure is at a potential of zero (i.e., it is at "ground" potential). Figure shows the geometry of the problem. Because the situation is static, there is no electric field inside the material of the conductors. If there were such a field, the charges that are free to move in a conducting material would do so until equilibrium was reached. The charges are arranged so that their individual contributions to the electric field at points inside the conducting material add up to zero. In a situation of static equilibrium, excess charges are located on the surface of conductors. Because there are no electric fields inside the conducting material, all parts of a given conductor are at the same potential; hence, a conductor is an equipotential in a static situation.

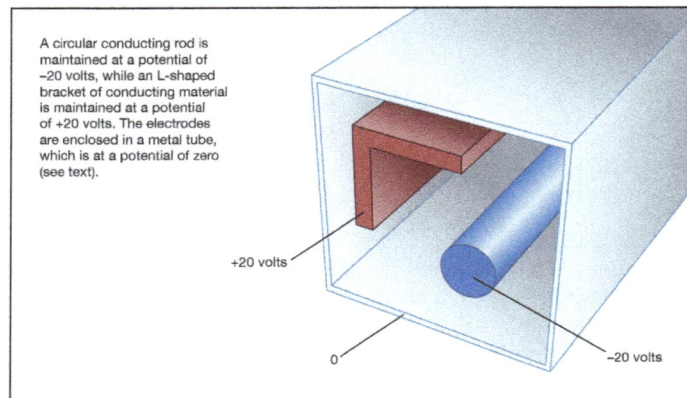

A circular conducting rod is maintained at a potential of −20 volts, while an L-shaped bracket of conducting material is maintained at a potential of +20 volts. The electrodes are enclosed in a metal tube, which is at a potential of zero (see text).

+20 volts

0

−20 volts

Electrode configuration.

In figure, the numerical solution of the problem gives the potential at a large number of points inside the cavity. The locations of the +20-volt and −20-volt electrodes can be recognized easily. In carrying out the numerical solution of the electrostatic problem in the figure, the electrostatic potential was determined directly by means of one of its important properties: in a region where there is no charge (in this case, between the conductors), the value of the potential at a given point is the average of the values of the potential in the neighbourhood of the point. This follows from the fact that the electrostatic potential in a charge-free region obeys Laplace's equation, which in vector calculus notation is div grad $V = 0$. This equation is a special case of Poisson's equation div grad $V = \rho$, which is applicable to electrostatic problems in regions where the volume charge density is ρ. Laplace's equation states that the divergence of the gradient of the potential is zero in regions of space with no charge. In the example of figure, the potential on the conductors remains constant. Arbitrary values of potential are initially assigned elsewhere inside the cavity. To obtain a solution, a computer replaces the potential at each coordinate point that is not on a conductor by the average of the values of the potential around that point; it scans the entire set of points many times until the values of the potentials differ by an amount small enough to indicate a satisfactory solution. Clearly, the larger the number of points, the more accurate the solution will be. The computation time as well as the computer memory size requirement increase rapidly, however, especially in three-dimensional problems with complex geometry. This method of solution is called the "relaxation" method.

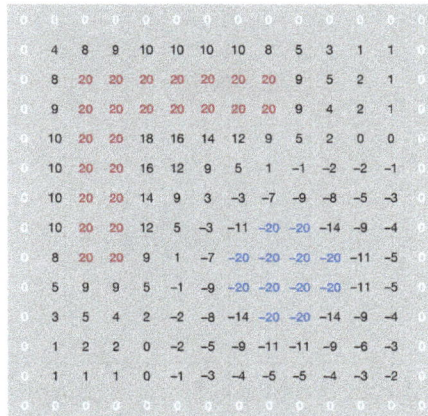

Numerical solution for the electrode configuration shown in Figure.
The electrostatic potentials are in volts.

In figure, points with the same value of electric potential have been connected to reveal a number of important properties associated with conductors in static situations. The lines in the figure represent equipotential surfaces. The distance between two equipotential surfaces tells how rapidly the potential changes, with the smallest distances corresponding to the location of the greatest rate of change and thus to the largest values of the electric field. Looking at the +20-volt and +15-volt equipotential surfaces, one observes immediately that they are closest to each other at the sharp external corners of the right-angle conductor. This shows that the strongest electric fields on the surface of a charged conductor are found on the sharpest external parts of the conductor; electrical breakdowns are most likely to occur there. It also should be noted that the electric field is weakest in the inside corners, both on the inside corner of the right-angle piece and on the inside corners of the square enclosure.

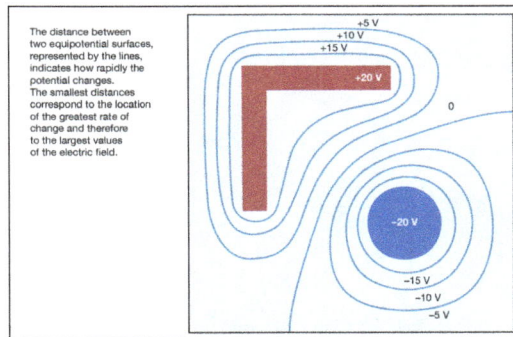

Equipotential surfaces.

In figure, dashed lines indicate the direction of the electric field. The strength of the field is reflected by the density of these dashed lines. Again, it can be seen that the field is strongest on outside corners of the charged L-shaped conductor; the largest surface charge density must occur at those locations. The field is weakest in the inside corners. The signs of the charges on the conducting surfaces can be deduced from the fact that electric fields point away from positive charges and toward negative charges. The magnitude of the surface charge density σ on the conductors is measured in coulombs per metre squared and is given by,

$$\sigma = \varepsilon_0 E,$$

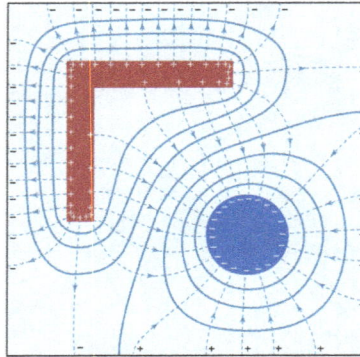

Electric field lines: The density of the dashed lines indicates the strength of the field.

where, ε_o is called the permittivity of free space and has the value of 8.854×10^{-12} coulomb squared per newton-square metre. In addition, ε_o is related to the constant k in Coulomb's law by,

$$k = \frac{1}{4\pi\varepsilon_0}.$$

Figure also illustrates an important property of an electric field in static situations: field lines are always perpendicular to equipotential surfaces. The field lines meet the surfaces of the conductors at right angles, since these surfaces also are equipotentials. Figure completes this example by showing the potential energy landscape of a small positive charge q in the region. From the variation in potential energy, it is easy to picture how electric forces tend to drive the positive charge q from higher to lower potential—i.e., from the L-shaped bracket at +20 volts toward the square-shaped enclosure at ground (0 volts) or toward the cylindrical rod maintained at a potential of −20 volts. It also graphically displays the strength of force near the sharp corners of conducting electrodes.

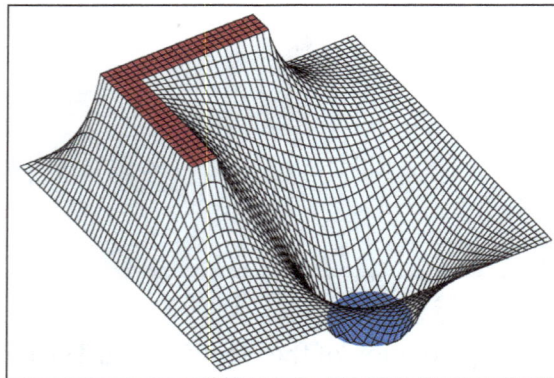

Potential energy for a positive charge.

Capacitance

A useful device for storing electrical energy consists of two conductors in close proximity and insulated from each other. A simple example of such a storage device is the parallel-plate capacitor. If positive charges with total charge +Q are deposited on one of the conductors and an equal amount of negative charge −Q is deposited on the second conductor, the capacitor is said to have a charge

Q. As shown in Figure, it consists of two flat conducting plates, each of area A, parallel to each other and separated by a distance d.

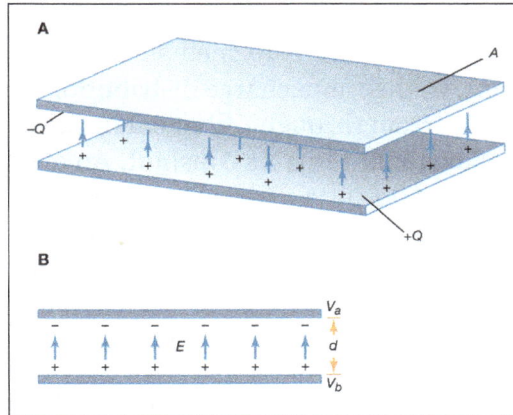

The links and boxes on the right side of the pageParallel-plate capacitor. (A) This storage device consists of two flat conducting plates, each of area A. (B) These plates are parallel and separated by a small distance d.

Electric Current

Current is a flow of electrical charge carriers, usually electrons or electron-deficient atoms. The common symbol for current is the uppercase letter I. The standard unit is the ampere, symbolized by A. One ampere of current represents one coulomb of electrical charge (6.24×10^{18} charge carriers) moving past a specific point in one second. Physicists consider current to flow from relatively positive points to relatively negative points; this is called conventional current or Franklin current. Electrons, the most common charge carriers, are negatively charged. They flow from relatively negative points to relatively positive points.

Electric current can be either direct or alternating. Direct current (DC) flows in the same direction at all points in time, although the instantaneous magnitude of the current might vary. In an alternating current (AC), the flow of charge carriers reverses direction periodically. The number of complete AC cycles per second is the frequency, which is measured in hertz. An example of pure DC is the current produced by an electrochemical cell. The output of a power-supply rectifier, prior to filtering, is an example of pulsating DC. The output of common utility outlets is AC.

Current per unit cross-sectional area is known as current density. It is expressed in amperes per square meter, amperes per square centimeter, or amperes per square millimeter. Current density can also be expressed in amperes per circular mil. In general, the greater the current in a conductor, the higher the current density. However, in some situations, current density varies in different parts of an electrical conductor. A classic example is the so-called skin effect, in which current density is high near the outer surface of a conductor, and low near the center. This effect occurs with alternating currents at high frequencies. Another example is the current inside an active electronic component such as a field-effect transistor (FET).

An electric current always produces a magnetic field. The stronger the current, the more intense the magnetic field. A pulsating DC, or an AC, characteristically produces an electromagnetic field. This is the principle by which wireless signal propagation occurs.

Direct Electric Current

Basic Phenomena and Principles

Many electric phenomena occur under what is termed steady-state conditions. This means that such electric quantities as current, voltage, and charge distributions are not affected by the passage of time. For instance, because the current through a filament inside a car headlight does not change with time, the brightness of the headlight remains constant. An example of a nonsteady-state situation is the flow of charge between two conductors that are connected by a thin conducting wire and that initially have an equal but opposite charge. As current flows from the positively charged conductor to the negatively charged one, the charges on both conductors decrease with time, as does the potential difference between the conductors. The current therefore also decreases with time and eventually ceases when the conductors are discharged.

In an electric circuit under steady-state conditions, the flow of charge does not change with time and the charge distribution stays the same. Since charge flows from one location to another, there must be some mechanism to keep the charge distribution constant. In turn, the values of the electric potentials remain unaltered with time. Any device capable of keeping the potentials of electrodes unchanged as charge flows from one electrode to another is called a source of electromotive force, or simply an emf.

Figure shows a wire made of a conducting material such as copper. By some external means, an electric field is established inside the wire in a direction along its length. The electrons that are free to move will gain some speed. Since they have a negative charge, they move in the direction opposite that of the electric field. The current i is defined to have a positive value in the direction of flow of positive charges. If the moving charges that constitute the current i in a wire are electrons, the current is a positive number when it is in a direction opposite to the motion of the negatively charged electrons. (If the direction of motion of the electrons were also chosen to be the direction of a current, the current would have a negative value.) The current is the amount of charge crossing a plane transverse to the wire per unit time—i.e., in a period of one second. If there are n free particles of charge q per unit volume with average velocity v and the cross-sectional area of the wire is A, the current i, in elementary calculus notation is,

$$i = \frac{dQ}{dt} = nevA,$$

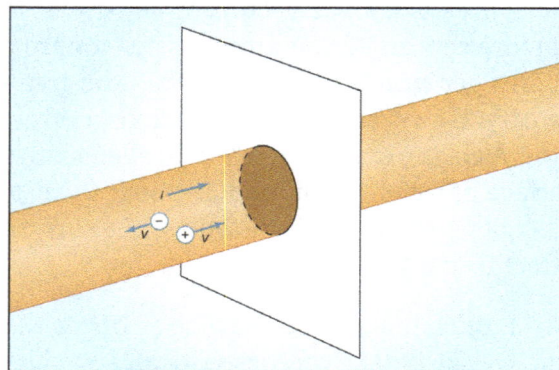

Motion of charge in electric current i.

where, dQ is the amount of charge that crosses the plane in a time interval dt. The unit of current is the ampere (A); one ampere equals one coulomb per second. A useful quantity related to the flow of charge is current density, the flow of current per unit area. Symbolized by J, it has a magnitude of i/A and is measured in amperes per square metre.

Wires of different materials have different current densities for a given value of the electric field E; for many materials, the current density is directly proportional to the electric field. This behaviour is represented by Ohm's law:

$$J = \sigma_J E.$$

The proportionality constant σ_J is the conductivity of the material. In a metallic conductor, the charge carriers are electrons and, under the influence of an external electric field, they acquire some average drift velocity in the direction opposite the field. In conductors of this variety, the drift velocity is limited by collisions, which heat the conductor.

If the wire in figure has a length l and area A and if an electric potential difference of V is maintained between the ends of the wire, a current i will flow in the wire. The electric field E in the wire has a magnitude V/l. The equation for the current, using Ohm's law, is:

$$i = JA = \frac{\sigma_J V}{l} A$$

Or

$$V = i\frac{l}{\sigma_J A}.$$

The quantity $l/\sigma_J A$, which depends on both the shape and material of the wire, is called the resistance R of the wire. Resistance is measured in ohms (Ω). The equation for resistance,

$$R = \frac{l}{\sigma_J A},$$

is often written as,

$$R = \frac{\rho l}{A},$$

where ρ is the resistivity of the material and is simply $1/\sigma J$. The geometric aspects of resistance in equation $R = \frac{\rho l}{A}$ are easy to appreciate: the longer the wire, the greater the resistance to the flow of charge. A greater cross-sectional area results in a smaller resistance to the flow.

The resistive strain gauge is an important application of equation $R = \frac{\rho l}{A}$. Strain, $\delta l/l$, is the fractional change in the length of a body under stress, where δl is the change of length and l is the length. The strain gauge consists of a thin wire or narrow strip of a metallic conductor such as constantan, an alloy of nickel and copper. A strain changes the resistance because the length, area, and resistivity of the conductor change. In constantan, the fractional change in resistance $\delta R/R$ is directly proportional to the strain with a proportionality constant of approximately 2.

A common form of Ohm's law is:

$$V = iR,$$

where, V is the potential difference in volts between the two ends of an element with an electric resistance of R ohms and where i is the current through that element.

Table lists the resistivities of certain materials at room temperature. These values depend to some extent on temperature; therefore, in applications where the temperature is very different from room temperature, the proper values of resistivities must be used to calculate the resistance. As an example, equation $R = \dfrac{\rho l}{A}$ shows that a copper wire 59 metres long and with a cross-sectional area of one square millimetre has an electric resistance of one ohm at room temperature.

Electric resistivities (at room temperature)	
Material	Resistivity (ohm-metre)
Silver	1.6×10^{-8}
Copper	1.7×10^{-8}
Aluminum	2.7×10^{-8}
Carbon (graphite)	1.4×10^{-5}
Germanium	4.7×10^{-1}
Silicon	2×10^{3}
Carbon (diamond)	5×10^{12}
Polyethylene	1×10^{17}
Fused quartz	$>1 \times 10^{19}$

Conductors, Insulators and Semiconductors

Materials are classified as conductors, insulators, or semiconductors according to their electric conductivity. The classifications can be understood in atomic terms. Electrons in an atom can have only certain well-defined energies, and, depending on their energies, the electrons are said to occupy particular energy levels. In a typical atom with many electrons, the lower energy levels are filled, each with the number of electrons allowed by a quantum mechanical rule known as the Pauli Exclusion Principle. Depending on the element, the highest energy level to have electrons may or may not be completely full. If two atoms of some element are brought close enough together so that they interact, the two-atom system has two closely spaced levels for each level of the single atom. If 10 atoms interact, the 10-atom system will have a cluster of 10 levels corresponding to each single level of an individual atom. In a solid, the number of atoms and hence the number of levels is extremely large; most of the higher energy levels overlap in a continuous fashion except for certain energies in which there are no levels at all. Energy regions with levels are called energy bands, and regions that have no levels are referred to as band gaps.

The highest energy band occupied by electrons is the valence band. In a conductor, the valence band is partially filled, and since there are numerous empty levels, the electrons are free to move under the influence of an electric field; thus, in a metal the valence band is also the conduction band. In an insulator, electrons completely fill the valence band; and the gap between it and the next band, which is the conduction band, is large. The electrons cannot move under the influence of an electric field unless they are given enough energy to cross the large energy gap to the conduction band. In a semiconductor, the gap to the conduction band is smaller than in an insulator. At room temperature, the valence band is almost completely filled. A few electrons are missing from the valence band because they have acquired enough thermal energy to cross the band gap to the conduction band; as a result, they can move under the influence of an external electric field. The "holes" left behind in the valence band are mobile charge carriers but behave like positive charge carriers.

For many materials, including metals, resistance to the flow of charge tends to increase with temperature. For example, an increase of 5 °C (9 °F) increases the resistivity of copper by 2 percent. In contrast, the resistivity of insulators and especially of semiconductors such as silicon and germanium decreases rapidly with temperature; the increased thermal energy causes some of the electrons to populate levels in the conduction band where, influenced by an external electric field, they are free to move. The energy difference between the valence levels and the conduction band has a strong influence on the conductivity of these materials, with a smaller gap resulting in higher conduction at lower temperatures.

The values of electric resistivities listed an extremely large variation in the capability of different materials to conduct electricity. The principal reason for the large variation is the wide range in the availability and mobility of charge carriers within the materials. For example, has many extremely mobile carriers; each copper atom has approximately one free electron, which is highly mobile because of its small mass. An electrolyte, such as a saltwater solution, is not as good a conductor as copper. The sodium and chlorine ions in the solution provide the charge carriers. The large mass of each sodium and chlorine ion increases as other attracted ions cluster around them. As a result, the sodium and chlorine ions are far more difficult to move than the free electrons in copper. Pure water also is a conductor, although it is a poor one because only a very small fraction of the water molecules are dissociated into ions. The oxygen, nitrogen, and argon gases that make up the atmosphere are somewhat conductive because a few charge carriers form when the gases are ionized by radiation from radioactive elements on Earth as well as from extraterrestrial cosmic rays (i.e., high-speed atomic nuclei and electrons). Electrophoresis is an interesting application based on the mobility of particles suspended in an electrolytic solution. Different particles (proteins, for example) move in the same electric field at different speeds; the difference in speed can be used to separate the contents of the suspension.

A current flowing through a wire heats it. This familiar phenomenon occurs in the heating coils of an electric range or in the hot tungsten filament of an electric light bulb. This ohmic heating is the basis for the fuses used to protect electric circuits and prevent fires; if the current exceeds a certain value, a fuse, which is made of an alloy with a low melting point, melts and interrupts the flow of current. The power P dissipated in a resistance R through which current i flows is given by,

$$P = i^2 R,$$

where, P is in watts (one watt equals one joule per second), i is in amperes, and R is in ohms. According to Ohm's law, the potential difference V between the two ends of the resistor is given by V = iR, and so the power P can be expressed equivalently as,

$$P = iV = \frac{V^2}{R}.$$

In certain materials, however, the power dissipation that manifests itself as heat suddenly disappears if the conductor is cooled to a very low temperature. The disappearance of all resistance is a phenomenon known as superconductivity. Electrons acquire some average drift velocity v under the influence of an electric field in a wire. Normally the electrons, subjected to a force because of an electric field, accelerate and progressively acquire greater speed. Their velocity is, however, limited in a wire because they lose some of their acquired energy to the wire in collisions with other electrons and in collisions with atoms in the wire. The lost energy is either transferred to other electrons, which later radiate, or the wire becomes excited with tiny mechanical vibrations referred to as phonons. Both processes heat the material. The term phonon emphasizes the relationship of these vibrations to another mechanical vibration—namely, sound. In a superconductor, a complex quantum mechanical effect prevents these small losses of energy to the medium. The effect involves interactions between electrons and also those between electrons and the rest of the material. It can be visualized by considering the coupling of the electrons in pairs with opposite momenta; the motion of the paired electrons is such that no energy is given up to the medium in inelastic collisions or phonon excitations. One can imagine that an electron about to "collide" with and lose energy to the medium could end up instead colliding with its partner so that they exchange momentum without imparting any to the medium.

A superconducting material widely used in the construction of electromagnets is an alloy of niobium and titanium. This material must be cooled to a few degrees above absolute zero temperature, −263.66 °C (or 9.5 K), in order to exhibit the superconducting property. Such cooling requires the use of liquefied helium, which is rather costly. During the late 1980s, materials that exhibit superconducting properties at much higher temperatures were discovered. These temperatures are higher than the −196 °C of liquid nitrogen, making it possible to use the latter instead of liquid helium. Since liquid nitrogen is plentiful and cheap, such materials may provide great benefits in a wide variety of applications, ranging from electric power transmission to high-speed computing.

Electromotive Force

A 12-volt automobile battery can deliver current to a circuit such as that of a car radio for a considerable length of time, during which the potential difference between the terminals of the battery remains close to 12 volts. The battery must have a means of continuously replenishing the excess positive and negative charges that are located on the respective terminals and that are responsible for the 12-volt potential difference between the terminals. The charges must be transported from one terminal to the other in a direction opposite to the electric force on the charges between the terminals. Any device that accomplishes this transport of charge constitutes a source of electromotive force. A car battery, for example, uses chemical reactions to generate electromotive force. The Van de Graaff generator shown in figure is a mechanical device that produces an electromotive force. Invented by the American physicist Robert J. Van de Graaff in the 1930s, this type of particle accelerator has been widely used to study subatomic particles.

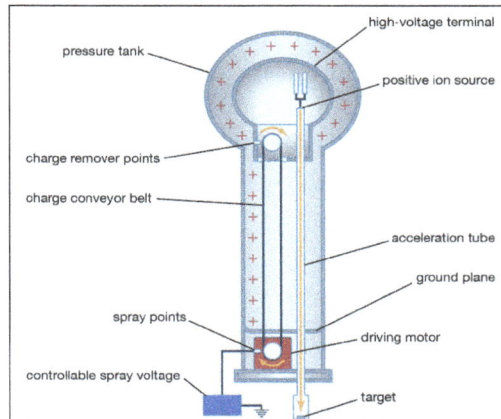

Van de Graaff accelerator

An insulating conveyor belt carries positive charge from the base of the Van de Graaff machine to the inside of a large conducting dome. The charge is removed from the belt by the proximity of sharp metal electrodes called charge remover points. The charge then moves rapidly to the outside of the conducting dome. The positively charged dome creates an electric field, which points away from the dome and provides a repelling action on additional positive charges transported on the belt toward the dome. Thus, work is done to keep the conveyor belt turning. If a current is allowed to flow from the dome to ground and if an equal current is provided by the transport of charge on the insulating belt, equilibrium is established and the potential of the dome remains at a constant positive value. In this example, the current from the dome to ground consists of a stream of positive ions inside the accelerating tube, moving in the direction of the electric field. The motion of the charge on the belt is in a direction opposite to the force that the electric field of the dome exerts on the charge. This motion of charge in a direction opposite the electric field is a feature common to all sources of electromotive force.

In the case of a chemically generated electromotive force, chemical reactions release energy. If these reactions take place with chemicals in close proximity to each other (e.g., if they mix), the energy released heats the mixture. To produce a voltaic cell, these reactions must occur in separate locations. A copper wire and a zinc wire poked into a lemon make up a simple voltaic cell. The potential difference between the copper and the zinc wires can be measured easily and is found to be 1.1 volts; the copper wire acts as the positive terminal. Such a "lemon battery" is a rather poor voltaic cell capable of supplying only small amounts of electric power. Another kind of 1.1-volt battery constructed with essentially the same materials can provide much more electricity. In this case, a copper wire is placed in a solution of copper sulfate and a zinc wire in a solution of zinc sulfate; the two solutions are connected electrically by a potassium chloride salt bridge. (A salt bridge is a conductor with ions as charge carriers.) In both kinds of batteries, the energy comes from the difference in the degree of binding between the electrons in copper and those in zinc. Energy is gained when copper ions from the copper sulfate solution are deposited on the copper electrode as neutral copper ions, thus removing free electrons from the copper wire. At the same time, zinc atoms from the zinc wire go into solution as positively charged zinc ions, leaving the zinc wire with excess free electrons. The result is a positively charged copper wire and a negatively charged zinc wire. The two reactions are separated physically, with the salt bridge completing the internal circuit.

Figure illustrates a 12-volt lead-acid battery, using standard symbols for depicting batteries in a circuit. The battery consists of six voltaic cells, each with an electromotive force of approximately two volts; the cells are connected in series, so that the six individual voltages add up to about 12 volts. As shown in figure, each two-volt cell consists of a number of positive and negative electrodes connected electrically in parallel. The parallel connection is made to provide a large surface area of electrodes, on which chemical reactions can take place. The higher rate at which the materials of the electrodes are able to undergo chemical transformations allows the battery to deliver a larger current.

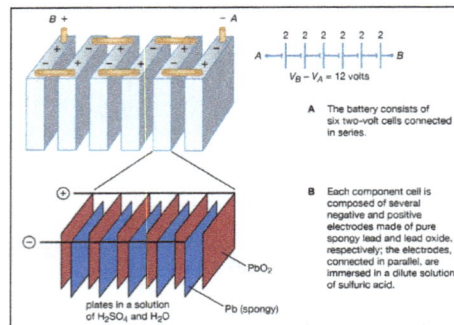

Voltaic cells and electrodes of a 12-volt lead-acid battery.

In the lead-acid battery, each voltaic cell consists of a negative electrode of pure, spongy lead (Pb) and a positive electrode of lead oxide (PbO_2). Both the lead and lead oxide are in a solution of sulfuric acid (H_2SO_4) and water (H_2O). At the positive electrode, the chemical reaction is $PbO_2 + SO^-/_4{}^- + 4H^+ + 2e^- \rightarrow PbSO_4 + 2H_2O + (1.68\ V)$. At the negative terminal, the reaction is $Pb + SO^-/_4{}^- \rightarrow PbSO_4 + 2e^- + (0.36\ V)$. The cell potential is $1.68 + 0.36 = 2.04$ volts. The 1.68 and 0.36 volts in the above equations are, respectively, the reduction and oxidation potentials; they are related to the binding of the electrons in the chemicals. When the battery is recharged, either by a car generator or by an external power source, the two chemical reactions are reversed.

Direct-current Circuits

The simplest direct-current (DC) circuit consists of a resistor connected across a source of electromotive force. The symbol for a resistor is shown in Figure; here the value of R, 60Ω, is given by the numerical value adjacent to the symbol. The symbol for a source of electromotive force, E, is shown with the associated value of the voltage. Convention gives the terminal with the long line a higher (i.e., more positive) potential than the terminal with the short line. Straight lines connecting various elements in a circuit are assumed to have negligible resistance, so that there is no change in potential across these connections. The circuit shows a 12-volt electromotive force connected to a 60Ω resistor. The letters a, b, c, and d on the diagram are reference points.

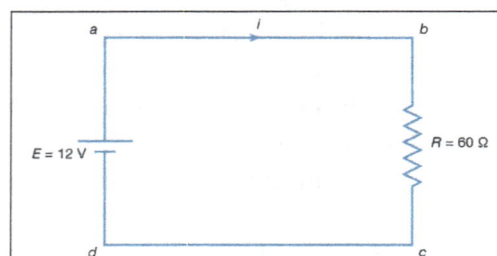

Direct-current circuit.

The function of the source of electromotive force is to maintain point a at a potential 12 volts more positive than point d. Thus, the potential difference Va − Vd is 12 volts. The potential difference across the resistance is Vb − Vc. From Ohm's law, the current i flowing through the resistor is:

$$i = \frac{V_b - V_c}{R} = \frac{V_b - V_c}{60}.$$

Since points a and b are connected by a conductor of negligible resistance, they are at the same potential. For the same reason, c and d are at the same potential. Therefore, $V_b - V_c = V_a - V_d =$ 12 volts. The current in the circuit is given by equation $i = \frac{V_b - V_c}{R} = \frac{V_b - V_c}{60}$. Thus, i = 12/60 = 0.2 ampere. The power dissipated in the resistor as heat is easily calculated using equation $P = i^2 R$:

$$P = i^2 R = (0.2)^2 \times 60 = 2.4 \ watts.$$

Where does the energy that is dissipated as heat in the resistor come from? It is provided by a source of electromotive force (e.g., a lead-acid battery). Within such a source, for each amount of charge dQ moved from the lower potential at d to the higher potential at a, an amount of work is done equal to dW = dQ(V_a − V_d). If this work is done in a time interval dt, the power delivered by the battery is obtained by dividing dW by dt. Thus, the power delivered by the battery (in watts) is:

$$\frac{dW}{dt} = (V_a - V_d)\frac{dQ}{dt} = (V_a - V_d)i.$$

Using the values i = 0.2 ampere and V_a − V_d = 12 volts makes dW/dt = 2.4 watts. As expected, the power delivered by the battery is equal to the power dissipated as heat in the resistor.

Resistors in Series and Parallel

If two resistors are connected in figure so that all of the electric charge must traverse both resistors in succession, the equivalent resistance to the flow of current is the sum of the resistances.

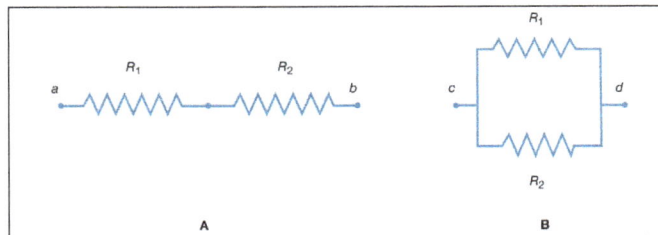

Resistors. (A) In series. (B) In parallel.

Using R_1 and R_2 for the individual resistances, the resistance between a and b is given by,

$$R_{ab} = R_1 + R_2$$

This result can be appreciated by thinking of the two resistors as two pieces of the same type of thin wire. Connecting the wires in series as shown simply increases their length to equal the sum of their two lengths. As equation $R = \frac{\rho l}{A}$ indicates, the resistance is the same as that given by equation $R_{ab} = R_1 + R_2$. The resistances R_1 and R_2 can be replaced in a circuit by the equivalent resistance Rab. If R_1 = 5Ω and R_2 = 2Ω, then Rab = 7Ω. If two resistors are connected as shown in Figure, the electric charges have alternate paths for flowing from c to d. The resistance to the flow of charge from c to d is clearly less than if either R_1 or R_2 were missing. Anyone who has ever had to find a way out of a crowded theatre can appreciate how much easier it is to leave a building with several exits than one with a single exit. The value of the equivalent resistance for two resistors in parallel is given by the equation,

$$\frac{1}{R_{cd}} = \frac{1}{R_1} + \frac{1}{R_2}$$

This relationship follows directly from the definition of resistance in equation $R = \frac{\rho l}{A}$, where 1/R is proportional to the area. If the resistors R_1 and R_2 are imagined to be wires of the same length and material, they would be wires with different cross-sectional areas. Connecting them in parallel is equivalent to placing them side by side, increasing the total area available for the flow of charge. Clearly, the equivalent resistance is smaller than the resistance of either resistor individually. As a numerical example, for R_1 = 5Ω and R_2 = 2Ω, $1/R_{cd}$ = 1/5 + 1/2 = 0.7. Therefore, R_{cd} = 1/0.7 = 1.43Ω. As expected, the equivalent resistance of 1.43 ohms is smaller than either 2 ohms or 5 ohms. It should be noted that both equations $R_{ab} = R_1 + R_2$ and $\frac{1}{R_{cd}} = \frac{1}{R_1} + \frac{1}{R_2}$ are given in a form in which they can be extended easily to any number of resistances.

Kirchhoff's Laws of Electric Circuits

Two simple relationships can be used to determine the value of currents in circuits. They are useful even in rather complex situations such as circuits with multiple loops. The first relationship deals with currents at a junction of conductors. Figure shows three such junctions, with the currents assumed to flow in the directions indicated.

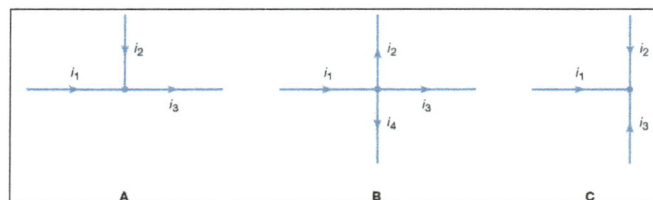

Electric currents at a junction.

Simply stated, the sum of currents entering a junction equals the sum of currents leaving that junction. This statement is commonly called Kirchhoff's first law (after the German physicist Gustav Robert Kirchhoff, who formulated it). For figure, the sum is $i_1 + i_2 = i_3$. For figure, $i_1 = i_2 + i_3 + i_4$. For Figure, $i_1 + i_2 + i_3$ = 0. If this last equation seems puzzling because all the currents appear to flow in and none flows out, it is because of the choice of directions for the individual currents. In solving a problem, the direction chosen for the currents is arbitrary. Once the problem has been solved, some currents have a positive value, and the direction arbitrarily chosen is the one of the actual current. In the solution some currents may have a negative value, in which case the actual current flows in a direction opposite that of the arbitrary initial choice.

Kirchhoff's second law is as follows: the sum of electromotive forces in a loop equals the sum of potential drops in the loop. When electromotive forces in a circuit are symbolized as circuit components as in Figure, this law can be stated quite simply: the sum of the potential differences across all the components in a closed loop equals zero. To illustrate and clarify this relation, one can consider a single circuit with two sources of electromotive forces E_1 and E_2, and two resistances R_1 and R_2, as shown in figure. The direction chosen for the current i also is indicated. The letters a, b, c, and d are used to indicate certain locations around the circuit. Applying Kirchhoff's second law to the circuit,

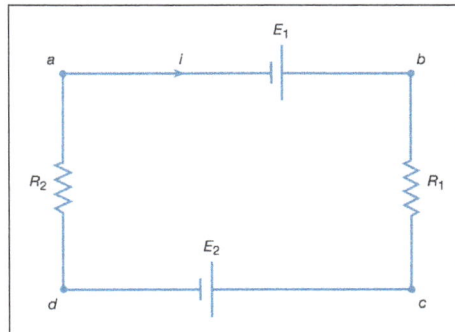

Circuit illustrating Kirchhoff's loop equation.

$$\left(V_b - V_a\right) + \left(V_c - V_b\right) + \left(V_d - V_c\right) + \left(V_a + V_d\right) = 0.$$

Referring to the circuit in figure, the potential differences maintained by the electromotive forces indicated are $V_b - V_a = E_1$, and $V_c - V_d = -E_2$. From Ohm's law, $V_b - V_c = iR_1$, and $V_d - V_a = iR_2$. Using these four relationships in equation $\left(V_b - V_a\right) + \left(V_c - V_b\right) + \left(V_d - V_c\right) + \left(V_a + V_d\right) = 0$, the so-called loop equation becomes $E_1 - E_2 - iR_1 - iR_2 = 0$.

Given the values of the resistances R_1 and R_2 in ohms and of the electromotive forces E_1 and E_2 in volts, the value of the current i in the circuit is obtained. If E_2 in the circuit had a greater value than E_1, the solution for the current i would be a negative value for i. This negative sign indicates that the current in the circuit would flow in a direction opposite the one indicated in figure.

Kirchhoff's laws can be applied to circuits with several connected loops. The same rules apply, though the algebra required becomes rather tedious as the circuits increase in complexity.

Alternating Electric Currents

Basic Phenomena and Principles

Many applications of electricity and magnetism involve voltages that vary in time. Electric power transmitted over large distances from generating plants to users involves voltages that vary sinusoidally in time, at a frequency of 60 hertz (Hz) in the United States and Canada and 50 hertz in Europe. (One hertz equals one cycle per second.) This means that in the United States, for example, the current alternates its direction in the electric conducting wires so that each second it flows 60 times in one direction and 60 times in the opposite direction. Alternating currents (AC) are also used in radio and television transmissions. In an AM (amplitude-modulation) radio broadcast, electromagnetic waves with a frequency of around one million hertz are generated by currents of the same frequency flowing back and forth in the antenna of the station. The information transported by these waves is encoded in the rapid variation of the wave amplitude. When voices and music are broadcast, these variations correspond to the mechanical oscillations of the sound and have frequencies from 50 to 5,000 hertz. In an FM (frequency-modulation) system, which is used by both television and FM radio stations, audio information is contained in the rapid fluctuation of the frequency in a narrow range around the frequency of the carrier wave.

Circuits that can generate such oscillating currents are called oscillators; they include, in addition to transistors, such basic electrical components as resistors, capacitors, and inductors. resistors dissipate heat while carrying a current. Capacitors store energy in the form of an electric field in the volume between oppositely charged electrodes. Inductors are essentially coils of conducting wire; they store magnetic energy in the form of a magnetic field generated by the current in the coil. All three components provide some impedance to the flow of alternating currents. In the case of capacitors and inductors, the impedance depends on the frequency of the current. With resistors, impedance is independent of frequency and is simply the resistance. This is easily seen from Ohm's law, equation $V = iR$, when it is written as i = V/R. For a given voltage difference V between the ends of a resistor, the current varies inversely with the value of R. The greater the value R, the greater is the impedance to the flow of electric current.

Transient Response

Consider a circuit consisting of a capacitor and a resistor that are connected as shown in Figure. What will be the voltage at point b if the voltage at a is increased suddenly from V_a = 0 to V_a = +50 volts? Closing the switch produces such a voltage because it connects the positive terminal of a 50-volt battery to point a while the negative terminal is at ground (point c). Figure graphs this voltage V_a as a function of the time.

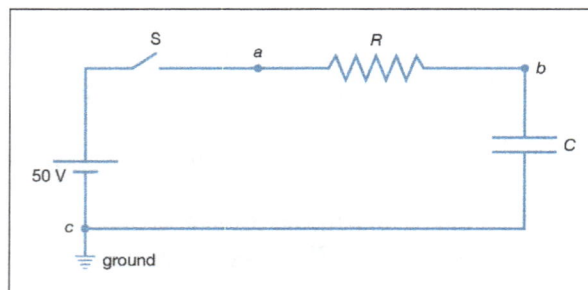

An RC circuit. This type of electric circuit consists of both a resistor and a capacitor connected as shown.

Initially, the capacitor has no charge and does not affect the flow of charge. The initial current is obtained from Ohm's law, V = iR, where V = V_a − V_b, V_a is 50 volts and V_b is zero. Using 2,000 ohms for the value of the resistance in Figure, there is an initial current of 25 milliamperes in the circuit. This current begins to charge the capacitor, so that a positive charge accumulates on the plate of the capacitor connected to point b and a negative charge accumulates on the other plate. As a result, the potential at point b increases from zero to a positive value. As more charge accumulates on the capacitor, this positive potential continues to increase. As it does so, the value of the potential across the resistor is reduced; consequently, the current decreases with time, approaching the value of zero as the capacitor potential reaches 50 volts. The behaviour of the potential at b in figure is described by the equation $V_b = V_a (1 - e^{-t/RC})$ in volts. For $R = 2,000\Omega$ and capacitance $C = 2.5$ microfarads, $V_b = 50(1 - e^{-t/0.005})$ in volts. The potential V_b at b in figure (right) increases from zero when the capacitor is uncharged and reaches the ultimate value of V_a when equilibrium is reached.

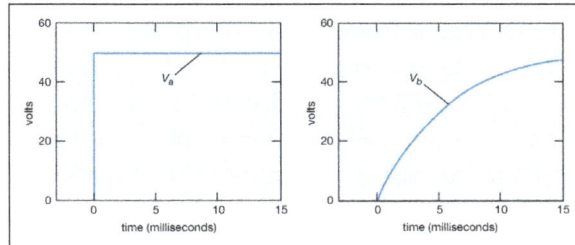

Voltage as a function of time.

How would the potential at point b vary if the potential at point a, instead of being maintained at +50 volts, were to remain at +50 volts for only a short time, say, one millisecond, and then return to zero? The superposition principle is used to solve the problem. The voltage at a starts at zero, goes to +50 volts at t = 0, then returns to zero at t = +0.001 second. This voltage can be viewed as the sum of two voltages, $V_1a + V_2a$, where V_1a becomes +50 volts at t = 0 and remains there indefinitely, and V_2a becomes −50 volts at t = 0.001 second and remains there indefinitely. This superposition is shown graphically on the left side of Figure. Since the solutions for V_1b and V_2b corresponding to V_1a and V_2a are known from the previous example, their sum V_b is the answer to the problem. The individual solutions and their sum are given graphically on the right side of figure.

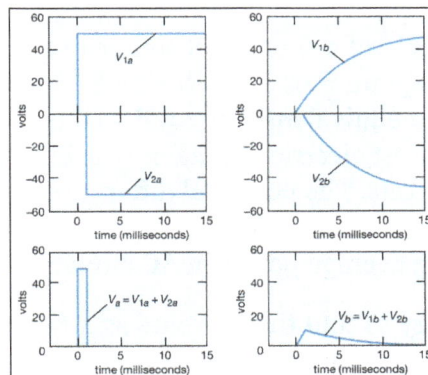

Application of the superposition principle to a problem concerned
with voltages as a function of time.

The voltage at b reaches a maximum of only 9 volts. The superposition illustrated in figure also shows that the shorter the duration of the positive "pulse" at a, the smaller is the value of the voltage generated at b. Increasing the size of the capacitor also decreases the maximum voltage at b. This decrease in the potential of a transient explains the "guardian role" that capacitors play in protecting delicate and complex electronic circuits from damage by large transient voltages. These transients, who generally occur at high frequency, produce effects similar to those produced by pulses of short duration. They can damage equipment when they induce circuit components to break down electrically. Transient voltages are often introduced into electronic circuits through power supplies. A concise way to describe the role of the capacitor in the above example is to say that its impedance to an electric signal decreases with increasing frequency. In the example, much of the signal is shunted to ground instead of appearing at point b.

Alternating-current Circuits

Certain circuits include sources of alternating electromotive forces of the sinusoidal form $V = V_0$ $\cos(\omega t)$ or $V = V_0 \sin(\omega t)$. The sine and cosine functions have values that vary between $+1$ and -1; either of the equations for the voltage represents a potential that varies with respect to time and has values from $+V_0$ to $-V_0$. The voltage varies with time at a rate given by the numerical value of ω; ω, which is called the angular frequency, is expressed in radians per second. Figure shows an example with $V_0 = 170$ volts and $\omega = 377$ radians per second, so that $V = 170 \cos(377t)$. The time interval required for the pattern to be repeated is called the period T, given by $T = 2\pi/\omega$. In figure 22, the pattern is repeated every 16.7 milliseconds, which is the period. The frequency of the voltage is symbolized by f and given by $f = 1/T$. In terms of ω, $f = \omega/2\pi$, in hertz.

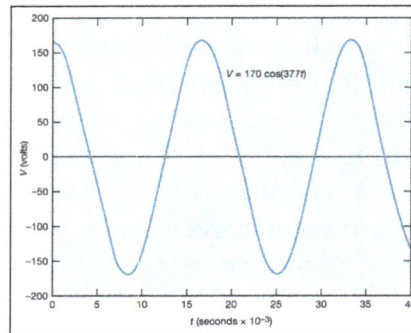

A sinusoidal voltage.

The root-mean-square (rms) voltage of a sinusoidal source of electromotive force (V_{rms}) is used to characterize the source. It is the square root of the time average of the voltage squared. The value of V_{rms} is V_0/Square root of$\sqrt{2}$, or, equivalently, $0.707V_0$. Thus, the 60-hertz, 120-volt alternating current, which is available from most electric outlets in U.S. homes and which is illustrated in figure 22, has $V_0 = 120/0.707 = 170$ volts. The potential difference at the outlet varies from $+170$ volts to -170 volts and back to $+170$ volts 60 times each second. The rms values of voltage and current are especially useful in calculating average power in AC circuits.

In AC circuits, it is often necessary to find the currents as a function of time in the various parts of the circuit for a given source of sinusoidal electromotive force. While the problems can become quite complex, the solutions are based on Kirchhoff's two laws. The solution for the current in a given loop takes the form $i = i_0 \cos(\omega t - \phi)$. The current has the same frequency as the applied voltage but is not necessarily "in phase" with that voltage. When the phase angle ϕ does not equal zero, the maximum of the current does not occur when the driving voltage is at its maximum.

Behaviour of an AC Circuit

The way an AC circuit functions can be better understood by examining one that includes a source of sinusoidally varying electromotive force, a resistor, a capacitor, and an inductor, all connected in series. For this single-loop problem, only the second of Kirchhoff's laws is needed since there is only one current. The circuit is shown in figure with the points a, b, c, and d at various positions in the circuit located between the various elements. The letters R, L, and C represent, respectively, the values of the resistance in ohms, the inductance in henrys, and the capacitance in farads. The source of the AC electromotive force is located between a and b. The wavy symbol is a reminder of the sinusoidal nature of the voltage that is responsible for making the current flow in the loop. For the potential between b and a,

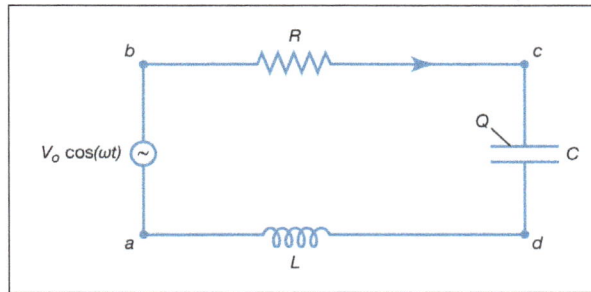

A series LRC circuit: This type of electric circuit has an inductor, resistor, and capacitor connected in series.

$$V_b - V_a = V_0 \cos \omega t.$$

Above equation represents a potential difference that has its maximum positive value at t = 0.

The direction chosen for the current i in the circuit in figure represent the direction of that current at some particular time, since AC circuits feature continuous reversals of the direction of the flow of charge. The direction chosen for the current is important, however, because the loop equation must consider all the elements at the same instant in time. The potential difference across the resistor is given by Ohm's law as,

$$V_b - V_c = iR.$$

For equation $V_b - V_c = iR$, the direction of the current is important. The potential difference across the capacitor, $V_c - V_d$, depends on the charge on the capacitor. When the charge on the upper plate of the capacitor in figure has a value Q, the potential difference across the capacitor is:

$$V_c = V_d = \frac{Q}{C}.$$

One must be careful labeling the charge and the direction of the current, since the charge on the other plate is −Q. For the choices shown in the figure, the current in the circuit is given by the rate of change of the charge Q—that is, i = dQ/dt. Finally, the value of the potential difference $V_d - V_a$ across the inductor depends on the rate of change of the current through the inductor, di/dt. For the direction chosen for i, the value is:

$$V_d - V_a = +L\frac{di}{dt}$$

The result of combining equations in accordance with Kirchhoff's second law for the loop in figure is:

$$V_0 \cos(\omega t) = L\frac{di}{dt} + iR + \frac{Q}{C}$$

Both the current i and the rate of change of the current di/dt can be eliminated from equation (28), since i = dQ/dt, and di/dt = d²Q/dt². The result is a linear, inhomogeneous, second-order differential equation with well-known solutions for the charge Q as a function of time. The most important solution describes the current and voltages after transient effects have been dampened; the transient effects last only a short time after the circuit is completed. Once the charge is known, the current in the circuit can be obtained by taking the first derivative of the charge. The expression for the current in the circuit is:

$$i = \frac{V_0}{Z}\cos(\omega t - \varphi) = i_0 \cos(\omega t - \varphi).$$

In above equation, Z is the impedance of the circuit; impedance, like resistance, is measured in units of ohms. Z is a function of the frequency of the source of applied electromotive force. The equation for Z is:

$$Z = \sqrt{R^2 + \left(\omega L - \frac{1}{\omega C}\right)^2}.$$

If the resistor were the only element in the circuit, the impedance would be Z = R, the resistance of the resistor. For a capacitor alone, Z = 1/ωC, showing that the impedance of a capacitor decreases as the frequency increases. For an inductor alone, Z = ωL Here it is sufficient to say that an induced electromotive force in the inductor opposes the change in current, and it is directly proportional to the frequency.

The phase angle ɸ in equation $i = \frac{V_0}{Z}\cos(\omega t - \varphi) = i_0 \cos(\omega t - \varphi)$ gives the time relationship between the current in the circuit and the driving electromotive force, $V_0 \cos(\omega t)$. The tangent of the angle ɸ is:

$$\tan\varphi = \frac{\left(\omega L - \dfrac{1}{\omega C}\right)}{R}.$$

Depending on the values of ω, L, and C, the angle ɸ can be positive, negative, or zero. If ɸ is positive, the current "lags" the voltage, while for negative values of ɸ, the current "leads" the voltage.

The power dissipated in the circuit is the same as the power delivered by the source of electromotive force, and both are measured in watts. Using equation $P = iV = \frac{V^2}{R}$, the power is given by,

$$P = iV = i_0 \cos(\omega t - \varphi)V_0 \cos(\omega t).$$

An expression for the average power dissipated in the circuit can be written either in terms of the peak values i_0 and V_0 or in terms of the rms values i_{rms} and V_{rms}. The average power is:

$$P_{ave} = I_{rms}V_{rms}\cos\varphi = \frac{1}{2}i_0 V_0 \cos\varphi.$$

The cos φ in above equation is called the power factor. It is evident that the only element that can dissipate energy is the resistance.

Resonance

A most interesting condition known as resonance occurs when the phase angle is zero in equation $\tan\varphi = \dfrac{\left(\omega L - \dfrac{1}{\omega C}\right)}{R}$, or equivalently, when the angular frequency ω has the value ω = ωr = Square root of√1/LC. The impedance in equation $Z = \sqrt{R^2 + \left(\omega L - \dfrac{1}{\omega C}\right)^2}$ then has its minimum value and equals the resistance R. The amplitude of the current in the circuit, i_o, is at its maximum value. Figure shows the dependence of i_o on the angular frequency ω of the source of alternating electromotive force. The values of the electric parameters for the figure are V_o = 50 volts, R = 25 ohms, L = 4.5 millihenrys, and C = 0.2 microfarad. With these values, the resonant angular frequency ωr of the circuit in figure is 3.33×10^4 radians per second.

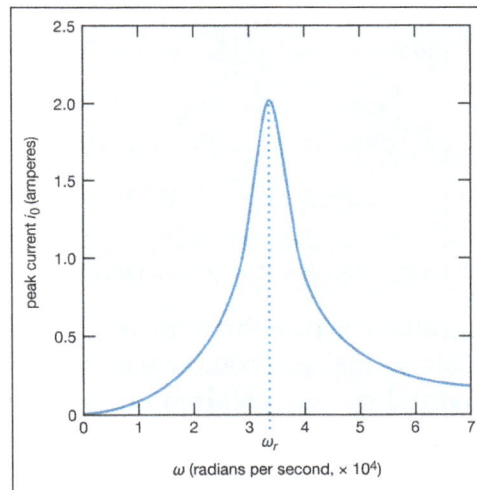

Current amplitude (peak current) as a function of ω.

The peaking in the current shown in figure constitutes a resonance. At the resonant frequency, in this case when ω_r equals 3.33×10^4 radians per second, the impedance Z of the circuit is at a minimum and the power dissipated is at a maximum. The phase angle φ is zero so that the current is in phase with the driving voltage, and the power factor, cos φ, is 1. Figure illustrates the variation of the average power with the angular frequency of the sinusoidal electromotive force. The resonance is seen to be even more pronounced. The quality factor Q for the circuit is the electric energy stored in the circuit divided by the energy dissipated in one period. The Q of a circuit is an important quantity in certain applications, as in the case of electromagnetic waveguides and radio-frequency cavities where Q has values around 10,000 and where high voltages and electric fields are desired. For the present circuit, Q = ωrL/R. Q also can be obtained from the average power graph as the ratio $\omega_r/(\omega_2 - \omega_1)$, where ω_1 and ω_2 are the angular frequencies at which the average power dissipated in the circuit is one-half its maximum value. For the circuit here, Q = 6.

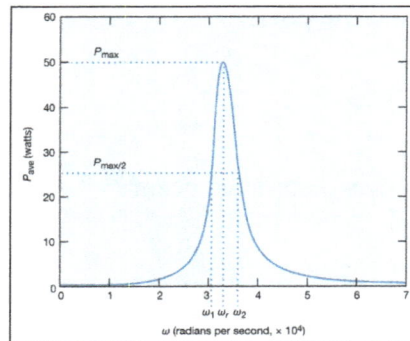

Average power dissipation versus ω.

What is the maximum value of the potential difference across the inductor? Since it is given by Ldi/dt, it will occur when the current has the maximum rate of change. Figure shows the amplitude of the potential difference as a function of ω.

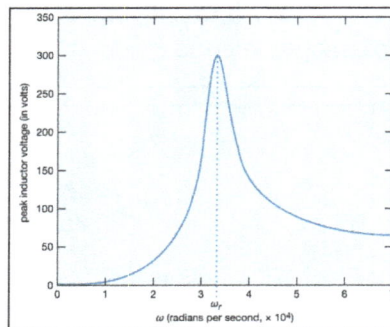

Electromotive force across L versus ω.

The maximum amplitude of the voltage across the inductor, 300 volts, is much greater than the 50-volt amplitude of the driving sinusoidal electromotive force. This result is typical of resonance phenomena. In a familiar mechanical system, children on swings time their kicks to attain very large swings (much larger than they could attain with a single kick). In a more spectacular, albeit costly, example, the collapse of the Tacoma Narrows Bridge (a suspension bridge across the Narrows of Puget Sound, Washington) on November 7, 1940, was the result of the large amplitudes of oscillations that the span attained as it was driven in resonance by high winds. A ubiquitous example of electric resonance occurs when a radio dial is turned to receive a broadcast. Turning the dial changes the value of the tuning capacitor of the radio. When the circuit attains a resonance frequency corresponding to the frequency of the radio wave, the voltage induced is enhanced and processed to produce sound.

Electric Properties of Matter

Piezoelectricity

Some solids, notably certain crystals, have permanent electric polarization. Other crystals become electrically polarized when subjected to stress. In electric polarization, the centre of positive charge within an atom, molecule, or crystal lattice element is separated slightly from the centre of negative charge. Piezoelectricity (literally "pressure electricity") is observed if a stress is applied to a solid, for example, by bending, twisting, or squeezing it. If a thin slice of quartz is compressed between two electrodes, a potential difference occurs; conversely, if the quartz crystal is inserted into an electric field, the resulting stress changes its dimensions. Piezoelectricity is responsible for the great precision of clocks and watches equipped with quartz oscillators. It also is used in electric guitars and various other musical instruments to transform mechanical vibrations into corresponding electric signals, which are then amplified and converted to sound by acoustical speakers.

A crystal under stress exhibits the direct piezoelectric effect; a polarization P, proportional to the stress, is produced. In the converse effect, an applied electric field produces a distortion of the crystal, represented by a strain proportional to the applied field. The basic equations of piezoelectricity are $P = d \times$ stress and $E =$ strain/d. The piezoelectric coefficient d (in metres per volt) is approximately 3×10^{-12} for quartz, 5×-10^{-11} for ammonium dihydrogen phosphate, and 3×10^{-10} for lead zirconate titanate.

For an elastic body, the stress is proportional to the strain—i.e., stress = $Y_e \times$ strain. The proportionality constant is the coefficient of elasticity Ye, also called Young's modulus for the English physicist Thomas Young. Using that relation, the induced polarization can be written as $P = dY_e \times$ strain, while the stress required to keep the strain constant when the crystal is in an electric field is stress = $-dY_eE$. The strain in a deformed elastic body is the fractional change in the dimensions of the body in various directions; the stress is the internal pressure along the various directions. Both are second-rank tensors, and, since electric field and polarization are vectors, the detailed treatment of piezoelectricity is complex. The equations above are oversimplified but can be used for crystals in certain orientations.

The polarization effects responsible for piezoelectricity arise from small displacements of ions in the crystal lattice. Such an effect is not found in crystals with a centre of symmetry. The direct effect can be quite strong; a potential $V = Yed\delta/\varepsilon_o K$ is generated in a crystal compressed by an amount δ, where K is the dielectric constant. If lead zirconate titanate is placed between two electrodes and a pressure causing a reduction of only 1/20th of one millimetre is applied, a 100,000-volt potential is produced. The direct effect is used, for example, to generate an electric spark with which to ignite natural gas in a heating unit or an outdoor cooking grill.

In practice, the converse piezoelectric effect, which occurs when an external electric field changes the dimensions of a crystal, is small because the electric fields that can be generated in a laboratory are minuscule compared to those existing naturally in matter. A static electric field of 10^6 volts per metre produces a change of only about 0.001 millimetre in the length of a one-centimetre quartz crystal. The effect can be enhanced by the application of an alternating electric field of the same frequency as the natural mechanical vibration frequency of the crystal. Many of the crystals have a quality factor Q of several hundred, and, in the case of quartz, the value can be 10^6. The result is a piezoelectric coefficient a factor Q higher than for a static electric field. The very large Q of quartz is exploited in electronic oscillator circuits to make remarkably accurate timepieces. The mechanical vibrations that can be induced in a crystal by the converse piezoelectric effect are also used to generate ultrasound, which is sound with a frequency far higher than frequencies audible to the human ear—above 20 kilohertz. The reflected sound is detectable by the direct effect. Such effects form the basis of ultrasound systems used to fathom the depths of lakes and waterways and to locate fish. Ultrasound has found application in medical imaging (e.g., fetal monitoring and the detection of abnormalities such as prostate tumours). The use of ultrasound makes it possible to produce detailed pictures of organs and other internal structures because of the variation in the reflection of sound from various body tissues. Thin films of polymeric plastic with a piezoelectric coefficient of about 10^{-11} metres per volt have been developed and have numerous applications as pressure transducers.

Electro-optic Phenomena

The index of refraction n of a transparent substance is related to its electric polarizability and is given by $n^2 = 1 + \chi e/\varepsilon_0$. χe is the electric susceptibility of a medium, and the equation $P = \chi eE$ relates the polarization of the medium to the applied electric field. For most matter, χe is not a constant independent of the value of the electric field, but rather depends to a small degree on the value of the field. Thus, the index of refraction can be changed by applying an external electric field to a medium. In liquids, glasses, and crystals that have a centre of symmetry, the change is usually very small. Called the Kerr effect (for its discoverer, the Scottish physicist John Kerr), it is proportional to the square of the applied electric field. In noncentrosymmetric crystals, the change in the index of refraction n is generally much greater; it depends linearly on the applied electric field and is known as the Pockels effect.

A varying electric field applied to a medium will modulate its index of refraction. This change in the index of refraction can be used to modulate light and make it carry information. A crystal widely used for its Pockels effect is potassium dihydrogen phosphate, which has good optical properties and low dielectric losses even at microwave frequencies.

An unusually large Kerr effect is found in nitrobenzene, a liquid with highly "acentric" molecules that have large electric dipole moments. Applying an external electric field partially aligns the otherwise randomly oriented dipole moments and greatly enhances the influence of the field on the index of refraction. The length of the path of light through nitrobenzene can be adjusted easily because it is a liquid.

Thermoelectricity

When two metals are placed in electric contact, electrons flow out of the one in which the electrons are less bound and into the other. The binding is measured by the location of the so-called Fermi level of electrons in the metal; the higher the level, the lower is the binding. The Fermi level represents the demarcation in energy within the conduction band of a metal between the energy levels occupied by electrons and those that are unoccupied. The energy of an electron at the Fermi level is $-W$ relative to a free electron outside the metal. The flow of electrons between the two conductors in contact continues until the change in electrostatic potential brings the Fermi levels of the two metals (W_1 and W_2) to the same value. This electrostatic potential is called the contact potential ϕ_{12} and is given by $e\phi_{12} = W_1 - W_2$, where e is 1.6×10^{-19} coulomb.

If a closed circuit is made of two different metals, there will be no net electromotive force in the circuit because the two contact potentials oppose each other and no current will flow. There will be a current if the temperature of one of the junctions is raised with respect to that of the second. There is a net electromotive force generated in the circuit, as it is unlikely that the two metals will have Fermi levels with identical temperature dependence. To maintain the temperature difference, heat must enter the hot junction and leave the cold junction; this is consistent with the fact that the current can be used to do mechanical work. The generation of a thermal electromotive force at a junction is called the Seebeck effect. The electromotive force is approximately linear with the temperature difference between two junctions of dissimilar metals, which are called a thermocouple. For a thermocouple made of iron and constantan (an alloy of 60 percent copper and 40 percent nickel), the electromotive force is about five millivolts when the cold junction is at 0 °C and the hot junction at 100 °C. One of the principal applications of the Seebeck effect is the measurement of temperature. The chemical properties of the medium, the temperature of which is measured, and the sensitivity required dictate the choice of components of a thermocouple.

The absorption or release of heat at a junction in which there is an electric current is called the Peltier effect. Both the Seebeck and Peltier effects also occur at the junction between a metal and a semiconductor and at the junction between two semiconductors. The development of semiconductor thermocouples (e.g., those consisting of n-type and p-type bismuth telluride) has made the use of the Peltier effect practical for refrigeration. Sets of such thermocouples are connected electrically in series and thermally in parallel. When an electric current is made to flow, a temperature difference, which depends on the current, develops between the two junctions. If the temperature of the hotter junction is kept low by removing heat, the second junction can be tens of degrees colder and act as a refrigerator. Peltier refrigerators are used to cool small bodies; they are compact, have no moving mechanical parts, and can be regulated to maintain precise and stable temperatures. They are employed in numerous applications, as, for example, to keep the temperature of a sample constant while it is on a microscope stage.

Thermionic Emission

A metal contains mobile electrons in a partially filled band of energy levels—i.e., the conduction band. These electrons, though mobile within the metal, are rather tightly bound to it. The energy that is required to release a mobile electron from the metal varies from about 1.5 to 6 electron volts, depending on the metal. In thermionic emission, some of the electrons acquire enough energy from thermal collisions to escape from the metal. The number of electrons emitted and therefore the thermionic emission current depend critically on temperature.

In a metal the conduction-band levels are filled up to the Fermi level, which lies at an energy −W relative to a free electron outside the metal. The work function of the metal, which is the energy required to remove an electron from the metal, is therefore equal to W. At a temperature of 1,000 K only a small fraction of the mobile electrons have sufficient energy to escape. The electrons that can escape are moving so fast in the metal and have such high kinetic energies that they are unaffected by the periodic potential caused by atoms of the metallic lattice. They behave like electrons trapped in a region of constant potential. Because of this, when the rate at which electrons escape from the metal is calculated, the detailed structure of the metal has little influence on the final result. A formula known as Richardson's law (first proposed by the English physicist Owen W. Richardson) is roughly valid for all metals. It is usually expressed in terms of the emission current density (J) as,

$$J = AT^2 e^{-W/kT}$$

in amperes per square metre. The Boltzmann constant k has the value 8.62×10^{-5} electron volts per kelvin, and temperature T is in kelvins. The constant A is 1.2×10^6 ampere degree squared per square metre, and varies slightly for different metals. For tungsten, which has a work function W of 4.5 electron volts, the value of A is 7×10^5 amperes per square metre kelvin squared and the current density at T equaling 2,400 K is 0.14 ampere per square centimetre. J rises rapidly with temperature. If T is increased to 2,600 K, J rises to 0.9 ampere per square centimetre. Tungsten does not emit appreciably at 2,000 K or below (less than 0.05 milliampere per square centimetre) because its work function of 4.5 electron volts is large compared to the thermal energy kT, which is only 0.16 electron volt. At 1,000 K, a mixture of barium and strontium oxides has a work function of approximately 1.3 electron volts and is a reasonably good conductor. Currents of several amperes per square centimetre can be drawn from such oxide cathodes, but in practice the current density is generally less than 0.2 ampere per square centimetre. The oxide layer deteriorates rapidly when higher current densities are drawn.

Secondary Electron Emission

If electrons with energies of 10 to 1,000 electron volts strike a metal surface in a vacuum, their energy is lost in collisions in a region near the surface, and most of it is transferred to other electrons in the metal. Because this occurs near the surface, some of these electrons may be ejected from the metal and form a secondary emission current. The ratio of secondary electrons to incident electrons is known as the secondary emission coefficient. For low-incident energies (below about one electron volt), the primary electrons tend to be reflected and the secondary emission coefficient is near unity. With increasing energy, the coefficient at first falls and then at about 10 electron volts begins to rise again, usually reaching a peak of value between 2 and 4 at energies of a few hundred electron volts. At higher energies, the primary electrons penetrate so far below the surface before losing energy that the excited electrons have little chance of reaching the surface and escaping. The secondary emission coefficients fall and, when the electrons have energies exceeding 20 kiloelectron volts, are usually well below unity. Secondary emission also can occur in insulators. Because many insulators have rather high secondary emission coefficients, it is often useful when high secondary emission yields are required to coat a metal electrode with a thin insulator layer a few atoms thick.

Photoelectric Conductivity

If light with a photon energy hv that exceeds the work function W falls on a metal surface, some of the incident photons will transfer their energy to electrons, which then will be ejected from the metal. Since hv is greater than W, the excess energy hv – W transferred to the electrons will be observed as their kinetic energy outside the metal. The relation between electron kinetic energy E and the frequency v (that is, E = hv – W) is known as the Einstein relation, and its experimental verification helped to establish the validity of quantum theory. The energy of the electrons depends on the frequency of the light, while the intensity of the light determines the rate of photoelectric emission.

In a semiconductor the valence band of energy levels is almost completely full while the conduction band is almost empty. The conductivity of the material derives from the few holes present in the valence band and the few electrons in the conduction band. Electrons can be excited from the valence to the conduction band by light photons having an energy hv that is larger than energy gap Eg between the bands. The process is an internal photoelectric effect. The value of Eg varies from semiconductor to semiconductor. For lead sulfide, the threshold frequency occurs in the infrared, whereas for zinc oxide it is in the ultraviolet. For silicon, eg., equals 1.1 electron volts, and the threshold wavelength is in the infrared, about 1,100 nanometres. Visible radiation produces electron transitions with almost unity quantum efficiency in silicon. Each transition yields a hole–electron pair (i.e., two carriers) that contributes to electric conductivity. For example, if one milliwatt of light strikes a sample of pure silicon in the form of a thin plate one square centimetre in area and 0.03 centimetre thick (which is thick enough to absorb all incident light), the resistance of the plate will be decreased by a factor of about 1,000. In practice, photoconductive effects are not usually as large as this, but this example indicates that appreciable changes in conductivity can occur even with low illumination. Photoconductive devices are simple to construct and are used to detect visible, infrared, and ultraviolet radiation.

Electroluminescence

Conduction electrons moving in a solid under the influence of an electric field usually lose kinetic energy in low-energy collisions as fast as they acquire it from the field. Under certain circumstances in semiconductors, however, they can acquire enough energy between collisions to excite atoms in the next collision and produce radiation as the atoms de-excite. A voltage applied across a thin layer of zinc sulfide powder causes just such an electroluminescent effect. Electroluminescent panels are of more interest as signal indicators and display devices than as a source of general illumination.

A somewhat similar effect occurs at the junction in a reverse-biased semiconductor p–n junction diode—i.e., a p–n junction diode in which the applied potential is in the direction of small current flow. Electrons in the intense field at the depleted junction easily acquire enough energy to excite atoms. Little of this energy finally emerges as light, though the effect is readily visible under a microscope.

When a junction between a heavily doped n-type material and a less doped p-type material is forward-biased so that a current will flow easily, the current consists mainly of electrons injected from the n-type material into the conduction band of the p-type material. These electrons ultimately drop into holes in the valence band and release energy equal to the energy gap of the material. In most cases, this energy E_g is dissipated as heat, but in gallium phosphide and especially in gallium arsenide, an appreciable fraction appears as radiation, the frequency ν of which satisfies the relation $h\nu = E_g$. In gallium arsenide, though up to 30 percent of the input electric energy is available as radiation, the characteristic wavelength of 900 nanometres is in the infrared. Gallium phosphide gives off visible green light but is inefficient; other related III-V compound semiconductors emit light of different colours. Electroluminescent injection diodes of such materials, commonly known as light-emitting diodes (LEDs), are employed mainly as indicator lamps and numeric displays. Semiconductor lasers built with layers of indium phosphide and of gallium indium arsenide phosphide have proved more useful. Unlike gas or optically pumped lasers, these semiconductor lasers can be modulated directly at high frequencies. They are used not only in devices such as compact disc players but also as light sources for long-distance optical fibre communications systems.

Bioelectric Effects

Bioelectricity refers to the generation or action of electric currents or voltages in biological processes. Bioelectric phenomena include fast signaling in nerves and the triggering of physical processes in muscles or glands. There is some similarity among the nerves, muscles, and glands of all organisms, possibly because fairly efficient electrochemical systems evolved early. Scientific studies tend to focus on the following: nerve or muscle tissue; such organs as the heart, brain, eye, ear, stomach, and certain glands; electric organs in some fish; and potentials associated with damaged tissue.

Electric activity in living tissue is a cellular phenomenon, dependent on the cell membrane. The membrane acts like a capacitor, storing energy as electrically charged ions on opposite sides of the membrane. The stored energy is available for rapid utilization and stabilizes the membrane system so that it is not activated by small disturbances.

Cells capable of electric activity show a resting potential in which their interiors are negative by about 0.1 volt or less compared with the outside of the cell. When the cell is activated, the resting potential may reverse suddenly in sign; as a result, the outside of the cell becomes negative and the inside positive. This condition lasts for a short time, after which the cell returns to its original resting state. This sequence, called depolarization and repolarization, is accompanied by a flow of substantial current through the active cell membrane, so that a "dipole-current source" exists for a short period. Small currents flow from this source through the aqueous medium containing the cell and are detectable at considerable distances from it. These currents, originating in active membrane, are functionally significant very close to their site of origin but must be considered incidental at any distance from it. In electric fish, however, adaptations have occurred, and this otherwise incidental electric current is actually utilized. In some species the external current is apparently used for sensing purposes, while in others it is used to stun or kill prey. In both cases, voltages from many cells add up in series, thus assuring that the specialized functions can be performed. Bioelectric potentials detected at some distance from the cells generating them may be as small as the 20 or 30 microvolts associated with certain components of the human electroencephalogram or the millivolt of the human electrocardiogram. On the other hand, electric eels can deliver electric shocks with voltages as large as 1,000 volts.

In addition to the potentials originating in nerve or muscle cells, relatively steady or slowly varying potentials (often designated dc) are known. These dc potentials occur in the following cases: in areas where cells have been damaged and where ionized potassium is leaking (as much as 50 millivolts); when one part of the brain is compared with another part (up to one millivolt); when different areas of the skin are compared (up to 10 millivolts); within pockets in active glands, e.g., follicles in the thyroid (as high as 60 millivolts); and in special structures in the inner ear (about 80 millivolts).

A small electric shock caused by static electricity during cold, dry weather is a familiar experience. While the sudden muscular reaction it engenders is sometimes unpleasant, it is usually harmless. Even though static potentials of several thousand volts are involved, a current exists for only a brief time and the total charge is very small. A steady current of two milliamperes through the body is barely noticeable. Severe electrical shock can occur above 10 milliamperes, however. Lethal current levels range from 100 to 200 milliamperes. Larger currents, which produce burns and unconsciousness, are not fatal if the victim is given prompt medical care. (Above 200 milliamperes, the heart is clamped during the shock and does not undergo ventricular fibrillation.) Prevention clearly includes avoiding contact with live electric wiring; risk of injury increases considerably if the skin is wet, as the electric resistance of wet skin may be hundreds of times smaller than that of dry skin.

Voltage

Voltage is one of the fundamental parameters associated with any electrical or electronic circuit. Voltage is seen widely in specifications of a host of electrical items from batteries to radios and light bulbs to shavers, and on top of this it is a key parameter that is measured within circuits as well.

The operating voltage of an item of equipment is very important - it is necessary to connect electrical and electronic items to supplies of the correct voltage. Connect a 240 volt light bulb to a 12 volt battery and it will not light up, but connect a small 5V USB device to a 240 volt supply and far too much current will flow and it will burn up and be irreparably damaged.

On top of this, the voltage levels within a circuit give a key to its operation - if the incorrect voltage is present, then it may give an indication of the reason for the malfunction.

For these and many reasons, electrical voltage is a key parameter and knowing what the voltage is can be a key requirement in any circumstance.

Voltage Basics

Voltage can be considered as the pressure that forces the charged electrons to flow in an electrical circuit. This flow of electrons is the electrical current that flows.

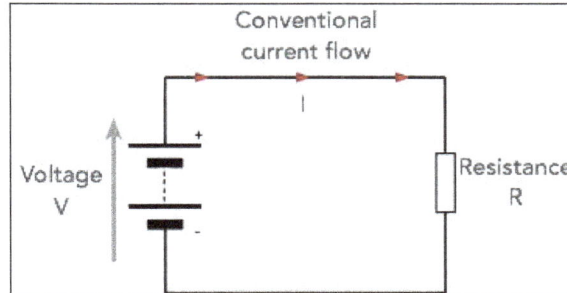

Voltage shown within a simple circuit

If a positive potential is placed on one end of a conductor, then this will attract that negative charges to it because unlike charges attract. The higher the potential attracting the charges, the greater the attraction and the greater the current flow.

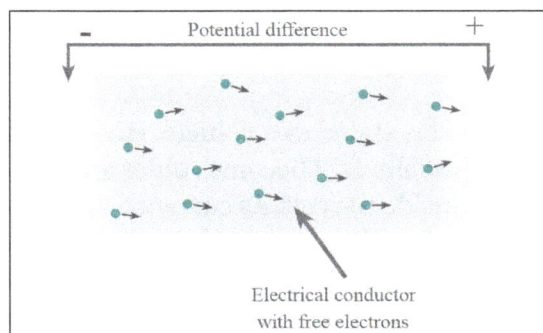

The higher the voltage potential difference, the greater the attraction of electrons and greater the current flow.

In essence, the voltage is the electrical pressure and it is measured in volts which can be represented by the letter V.

Normally the letter V is used for volts in an equation like Ohm's law, but occasionally the letter E may be used - this stands for EMF or electro-motive force.

To gain a view of what voltage is and how it affects electrical and electronic circuits, it is often useful as a basic analogy to think of water in a pipe, possibly even the plumbing system in a house. A water tank is placed up high to provide pressure (voltage) to force the water flow (current) through the pipes. The greater the pressure, the higher the water flow.

Potential Difference

The electrical potential or voltage is a measure of the electrical pressure available to force the current around a circuit. In the comparison of a water system mentioned when describing current, the potential can be likened to the water pressure at a given point. The greater the pressure difference across a section of the system, the greater the amount of water which will flow. Similarly the greater the potential difference or voltage across a section of an electrical circuit, the greater the current which will flow.

Volt

The basic unit of voltage is the volt, named after the Italian scientist, Alessandro Volta, who made some early batteries and performed many other experiments with electricity.

To give an idea of the voltages which are likely to be encountered, a CB radio will usually operate from a supply of around 12 volts (12 V). The cells used in domestic batteries have a voltage of around 1.5 volts. Rechargeable Nickel Cadmium cells have a slightly smaller voltage of 1.2 volts, but can normally be used interchangeably with the non-rechargeable types.

In other areas voltages much smaller and much greater than this can be encountered. The signal input to an audio amplifier will be smaller than this, and the voltages will often be measured in millivolts (mV) or thousandths of a volt. The signals at the input to a radio are even smaller than this and will often be measured in microvolts (μV) or millionths of a volt.

At the other extreme much greater voltages may be heard about. The cathode ray tubes in a television or computer monitors require voltages of several kilovolts (kV) or thousands of volts, and even larger voltages of millions of volts or megavolts (MV) may be heard of in conjunction with topics like lightning.

EMF and PD

The voltage for a battery or single cell is stated as a voltage. However it is found that when the battery is in use its voltage will fall, especially as it becomes older and it has been used. The reason for this is that there is some resistance inside the cell. As current flows a voltage drop forms across this and the voltage seen at the output is less than expected. Even so the voltage seen at the terminals if the battery was not supplying current would still be the same. This no load voltage is known as the electro-motive force (EMF), and is the internal voltage which is generated by the cell, or other source of power.

References

- Electrical-energy: circuitglobe.com, Retrieved 12 May, 2019

- Electric-power: circuitglobe.com, Retrieved 18 March, 2019

- Electric-power: sparkfun.com, Retrieved 8 July, 2019

- Electric-charge, electrostatics, electromagnetics: physics-and-radio-electronics.com, Retrieved 9 January, 2019

- Electricity, science: britannica.com, Retrieved 17 April, 2019

- Dielectrics-polarization-and-electric-dipole-moment, electricity, science: britannica.com, Retrieved 7 June, 2019

- What-is-voltage-basics-tutorial, voltage: electronics-notes.com, Retrieved 19 February, 2019

3

Basics of Electrical Circuits

Electrical circuits comprise of various components which are broadly categorized as active and passive components. The passive components are resistors, capacitors and inductors, and active components are diodes, transistors and integrated circuits. This chapter discusses in detail these components of electrical circuits.

Electrical Components

There are numerous important basic electrical components commonly found in the circuits of almost all peripherals. These devices are the essential building blocks of electronic and electrical circuits. These electric components can be found in great numbers on motherboards, video cards, hard disk, logic boards and everywhere else in personal computers. The electrical circuit components can be combined with each other and with dozens of other devices.

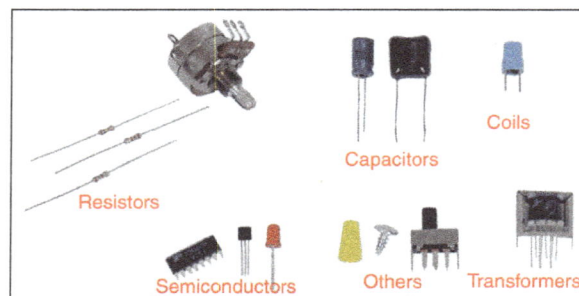

Electrical Components.

Electrical circuits on different parts of equipment share similarities. Many circuits have resistors, inductors, capacitors, transformers, fuses and switches.

Passive Components

In electrical systems, passive components are those that do not require electrical power to operate (e.g., not capable of power gain). This could include the capacitors, resistors or enclosures that do not require electrical power to operate but would exclude system components such as the transistors, diodes and other controllers.

Resistor

The resistor is a passive electrical component to create resistance in the flow of electric current. In almost all electrical networks and electronic circuits they can be found. The resistance is measured in ohms. An ohm is the resistance that occurs when a current of one ampere passes through a resistor with a one volt drop across its terminals. The current is proportional to the voltage across the terminal ends. This ratio is represented by Ohm's law:

$$R = \frac{V}{I}$$

Resistors are used for many purposes. A few examples include delimit electric current, voltage division, heat generation, matching and loading circuits, control gain, and fix time constants. They are commercially available with resistance values over a range of more than nine orders of magnitude. They can be used to as electric brakes to dissipate kinetic energy from trains, or be smaller than a square millimeter for electronics.

Resistor Types and Materials

Resistors can be divided in construction type as well as resistance material. The following breakdown for the type can be made:

- Fixed resistors

- Variable resistors, such as the:

  ○ Potentiometer

  ○ Rheostat

  ○ Trimpot

- Resistance dependent on a physical quantity:

  ○ Thermistors (NTC and PTC) as a result of temperature change.

  ○ Photo resistor (LDR) as a result of a changing light level.

  ○ Varistor (VDR) as a result of a changing voltage.

    ◦   Magneto resistor (MDR) as a result of a changing magnetic field.

    ◦   Strain Gauges as a result of mechanical load.

For each of these types a standard symbol exists. Another breakdown based on the material and manufacturing process can be made:

- Carbon composition

- Carbon film

- Metal film

- Metal oxide film

- Wirewound

The choice of material technology is a specific to the purpose. Often it is a trade-off between costs, precision and other requirements. For example, carbon composition is a very old technique with a low precision, but is still used for specific applications where high energy pulses occur. Carbon composition resistors have a body of a mixture of fine carbon particles and a non-conductive ceramic. The carbon film technique has a better tolerance. These are made of a non-conductive rod with a thin carbon film layer around it. This layer is treated with a spiral cut to increase and control the resistance value. Metal and metal oxide film are widely used nowadays, and have better properties for stability and tolerance. Furthermore, they are less influenced by temperature variations. They are just as carbon film resistors constructed with a resistive film around a cylindrical body. Metal oxide film is generally more durable. Wirewound resistors are probably the oldest type and can be used for both high precision as well as high power applications. They are constructed by winding a special metal alloy wire, such as nickel chrome, around a non-conductive core. They are durable, accurate and can have very low resistance value. A disadvantage is that they suffer from parasitic reactance at high frequencies. For the highest requirements on precision and stability, metal foil resistors are used. They are constructed by cementing a special alloy cold rolled film onto a ceramic substrate.

Resistor Characteristics

Dependent on the application, the electrical engineer specifies different properties of the resistor. The primary purpose is to limit the flow of electrical current; therefore the key parameter is the resistance value. The manufacturing accuracy of this value is indicated with the resistor tolerance in percentage. Many other parameters that affect the resistance value can be specified, such as long term stability or the temperature coefficient. The temperature coefficient, usually specified in high precision applications, is determined by the resistive material as well as the mechanical design.

In high frequency circuits, such as in radio electronics, the capacitance and inductance can lead to undesired effects. Foil resistors generally have a low parasitic reactance, while wirewound resistors are amongst the worst. For accurate applications such as audio amplifiers, the electric noise must be as low as possible. This is often specified as microvolts noise per volt of applied voltage, for a 1 MHz bandwidth. For high power applications the power rating is important. This specifies the maximum operating power the component can handle without altering the properties or damage. The power rating is usually specified in free air at room temperature. Higher power ratings require a larger size and may even require heat sinks. Many other characteristics can play a role in the design specification. Examples are the maximum voltage, or the pulse stability. In situations where high voltage surges could occur this is an important characteristic.

Sometimes not only the electrical properties are important, but the designer also has to consider the mechanical robustness in harsh environments. Military standards sometimes offer guidance to define the mechanical strength or the failure rate.

Resistor Standards

Many standards exist for resistors. The standards describe ways to measure and quantify important properties. Other norms exist for the physical size and resistance values. Probably, the most well known standard is the color code marking for axial leaded resistors.

Resistor Color Code

Resistor with a resistance of 5600 ohm with 2 % tolerance, according to the marking code IEC 60062.

The resistance value and tolerance are indicated with several colored bands around the component body. This marking technique of electronic components was already developed in the 1920's. Printing technology was still not far developed, what made printed numerical codes too difficult on small components. Nowadays, the color code is still used for most axial resistors up to one watt. In the figure an example is shown with four color bands. In this example the two first bands determine the significant digits of the resistance value, the third band is the multiplying factor and the fourth band gives the tolerance. Each color represents a different number and can be looked up in a resistor color code chart.

Resistor Color Code Calculator

The color code can easily be decoded using this calculator. It not only provides the resistance value, it also indicates when the value belongs to an E-series.

SMD Resistors

For SMD (Surface Mount Device) resistors a numerical code is used, because the components are too small for color coding. SMD resistors are -just as leaded variants – mainly available in the preferred values. The size of the component (length and width) is standardized as well, and is referred to as resistor package. The marking "331" means that the resistance has a value of $33\Omega \times 10^1 = 330\Omega$.

Resistor Values (Preferred values)

In the 1950s the increased production of resistors created the need for standardized resistance values. The range of resistance values is standardized with so called preferred values. The preferred values are defined in E-series. In an E-series, every value is a certain percentage higher than the previous. Various E-series exist for different tolerances.

Resistor Applications

There is a huge variation in fields of applications for resistors; from precision components in digital electronics, till measurement devices for physical quantities.

Resistors in Series and Parallel

In electronic circuits, resistors are very often connected in series or in parallel. A circuit designer might for example combine several resistors with standard values (E-series) to reach a specific resistance value. For series connection, the current through each resistor is the same and the equivalent resistance is equal to the sum of the individual resistors. For parallel connection, the voltage through each resistor is the same, and the inverse of the equivalent resistance is equal to the sum of the inverse values for all parallel resistors.

Measure Electrical Current (Shunt Resistor)

Electrical current can be calculated by measuring the voltage drop over a precision resistor with a known resistance, which is connected in series with the circuit. The current is calculated by using Ohm's law. This is a called an ammeter or shunt resistor. Usually this is a high precision manganin resistor with a low resistance value.

Resistors for LEDs

LED lights need a specific current to operate. A too low current will not light up the LED, while a too high current might burn out the device. Therefore, they are often connected in series with resistors. These are called ballast resistors and passively regulate the current in the circuit.

Blower Motor Resistor

In cars the air ventilation system is actuated by a fan that is driven by the blower motor. A special resistor is used to control the fan speed. This is called the blower motor resistor. Different designs are in use. One design is a series of different size wirewound resistors for each fan speed. Another design incorporates a fully integrated circuit on a printed circuit board.

Resistance

The electrical resistance is defined as the difficulty occurs in the flow of electrons. The conductor has free electrons. When the voltage or potential difference is applied across the conductor, the free electrons start moving in the particular direction. During the movement, these electrons collide with the atoms and molecules of the conductor. The atoms or molecules create the obstruction in the flow of electrons. This obstruction is called resistance.

The electrical resistance is provided to the circuit through the resistor. The resistance shows the relation between the applied voltage and the flow of charges. The resistance is inversely proportional to the flow of current.

Unit: Resistance is measured in ohms (kilo-ohms) and is denoted by symbols Ω (or kΩ). A wire is said to have a resistance of one ohm if the one-ampere current is passed through it.

Resistance Calculation

The resistance of a conductor is resistivity of the conductor's material times the conductor's length divided by the conductor's cross sectional area:

$$R = \rho \times \frac{l}{A}$$

Where,

R is the resistance in ohms (Ω).

ρ is the resistivity in ohms-meter ($\Omega \times$ m).

l is the length of the conductor in meter (m).

A is the cross sectional area of the conductor in square meters (m²).

It is easy to understand this formula with water pipes analogy:

- When the pipe is longer, the length is bigger and the resistance will increase.

- When the pipe is wider, the cross sectional area is bigger and the resistance will decrease.

Resistance Calculation with Ohm's Law

$$R = \frac{V}{I}$$

Where,

R is the resistance of the resistor in ohms (Ω).

V is the voltage drop on the resistor in volts (V).

I is the current of the resistor in amperes (A).

Temperature Effects of Resistance

The resistance of a resistor increases when temperature of the resistor increases:

$$R_2 = R_1 \times (1 + \alpha(T_2 - T_1))$$

Where,

R_2 is the resistance at temperature T_2 in ohms (Ω).

R_1 is the resistance at temperature T_1 in ohms (Ω).

α is the temperature coefficient.

Resistance of Resistors in Series

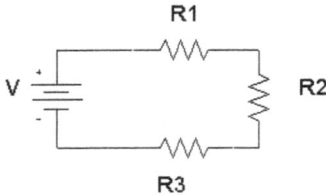

The total equivalent resistance of resistors in series is the sum of the resistance values:

$$R_{Total} = R_1 + R_2 + R_3 + ...$$

Resistance of Resistors in Parallel

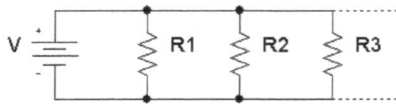

The total equivalent resistance of resistors in parallel is given by:

$$\frac{1}{R_{Total}} = \frac{1}{R_1} + \frac{1}{R_2} + \frac{1}{R_3} + ...$$

Measuring Electrical Resistance

- Electrical resistance is measured with ohmmeter instrument.

- In order to measure the resistance of a resistor or a circuit, the circuit should have the power supply turned off.

- The ohmmeter should be connected to the two ends of the circuit so the resistance can be read.

Resistivity or Coefficient of Resistance

Resistivity or Coefficient of Resistance is a property of substance, due to which the substance offers opposition to the flow of current through it. Resistivity or Coefficient of Resistance of any substance can easily be calculated from the formula derived from Laws of Resistance.

Laws of Resistance

The resistance of any substance depends on the following factors,

1. Length of the substance.

2. Cross sectional area of the substance.

3. The nature of material of the substance.

4. Temperature of the substance.

There are mainly four (4) laws of resistance from which the resistivity or specific resistance of any substance can easily be determined.

First Law of Resistance

The resistance of a substance is directly proportional to the length of the substance. electrical resistance R of a substance is:

$$R \propto L$$

Where L is the length of the substance.

If the length of a substance is increased, the path traveled by the electrons is also increased. If electrons travel long, they collide more and consequently the number of electrons passing through the substance becomes less; hence current through the substance is reduced. In other words, the resistance of the substance increases with increasing length of the substance. This relation is also linear.

Second Law of Resistance

The resistance of a substance is inversely proportional to the cross-sectional area of the substance. Electrical resistance R of a substance is:

$$R \propto \frac{1}{A}$$

Where A is the cross-sectional area of the substance.

The current through any substance depends on the numbers of electrons pass through a cross-section of substance per unit time. So, if the cross section of any substance is larger then more electrons can cross the cross section. Passing of more electrons through a cross-section per unit time causes more current through the substance. For fixed voltage, more current means less electrical resistance and this relation is linear.

Resistivity

Combining these two laws we get,

$$R \propto \frac{L}{A} \Rightarrow R = \rho \frac{L}{A}$$

Where, ρ (rho) is the proportionality constant and known as resistivity or specific resistance of the material of the conductor or substance. Now if we put, L = 1 and A = 1 in the equation, we get, R = ρ. That means resistance of a material of unit length having unit cross – sectional area is equal to its resistivity or specific resistance. Resistivity of a material can alternatively be defined as the electrical resistance between opposite faces of a cube of unit volume of that material.

Third Law of Resistance

The resistance of a substance is directly proportional to the resistivity of the materials by which the substance is made. The resistivity of all materials is not the same. It depends on the number of free electrons, and size of the atoms of the materials, types of bonding in the materials and many other factors of the material structures. If the resistivity of a material is high, the resistance offered by the substance made by this material is high and vice versa. This relation is also linear:

$$R \propto \rho$$

Fourth Law of Resistance

The temperature of the substance also affects the resistance offered by the substance. This is because, the heat energy causes more inter-atomic vibration in the metal, and hence electrons get more obstruction during drifting from lower potential end to higher potential end. Hence, in metallic substance, resistance increases with increasing temperature. If the substance in nonmetallic, with increasing temperature, the more covalent bonds are broken, these cause more free electrons in the material. Hence, resistance is decreased with increase in temperature.

That is why mentioning resistance of any substance without mentioning its temperature is meaningless.

Unit of Resistivity

The unit of resistivity can be easily determined form its equation:

$$R = \rho \frac{L}{A} \Rightarrow \rho = \frac{RA}{L}$$

In SI System of Unit

$$\rho = \frac{R\ \Omega \times A\ m^2}{L\ m}$$

$$\Rightarrow \rho = \frac{RA}{L}\frac{\Omega - m^2}{m}\ or\ \Omega - m$$

The unit of resistivity is $\Omega - m$ in MKS system and $\Omega - cm$ in CGS system and $1\ \Omega - m = 100\ \Omega - cm$.

Capacitor

A capacitor is a two-terminal, electrical component. Along with resistors and inductors, they are one of the most fundamental passive components we use.

What makes capacitors special is their ability to store energy; they're like a fully charged electric battery. *Caps*, as we usually refer to them, have all sorts of critical applications in circuits. Common applications include local energy storage, voltage spike suppression, and complex signal filtering.

Symbols and Units

Circuit Symbols

There are two common ways to draw a capacitor in a schematic. They always have two terminals, which go on to connect to the rest of the circuit. The capacitors symbol consists of two parallel lines, which are either flat or curved; both lines should be parallel to each other, close, but not touching (this is actually representative of how the capacitor is made. Hard to describe, easier to just show:

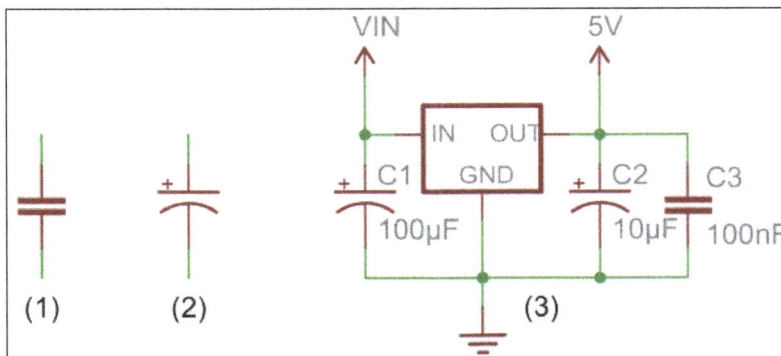

(1) and (2) are standard capacitor circuit symbols. (3) is an example of capacitors symbols in action in a voltage regulator circuit.

The symbol with the curved line indicates that the capacitor is polarized, meaning it's probably an electrolytic capacitor.

Each capacitor should be accompanied by a name - C1, C2, etc. and a value. The value should indicate the capacitance of the capacitor; how many farads it has.

Capacitance Units

Not all capacitors are created equal. Each capacitor is built to have a specific amount of capacitance. The capacitance of a capacitor tells you how much charge it can store, more capacitance means more capacity to store charge. The standard unit of capacitance is called the farad, which is abbreviated F.

It turns out that a farad is a lot of capacitance, even 0.001F (1 milifarad - 1mF) is a big capacitor. Usually you'll see capacitors rated in the pico- (10^{-12}) to microfarad (10^{-6}) range.

Prefix Name	Abbreviation	Weight	Equivalent Farads
Picofarad	pF	10^{-12}	0.000000000001 F
Nanofarad	nF	10^{-9}	0.000000001 F
Microfarad	µF	10^{-6}	0.000001 F
Milifarad	mF	10^{-3}	0.001 F
Kilofarad	kF	10^{3}	1000 F

When you get into the farad to kilofarad range of capacitance, you start talking about special caps called super or ultra-capacitors.

Capacitor Theory

The schematic symbol for a capacitor actually closely resembles how it's made. A capacitor is created out of two metal plates and an insulating material called a dielectric. The metal plates are placed very close to each other, in parallel, but the dielectric sits between them to make sure they don't touch.

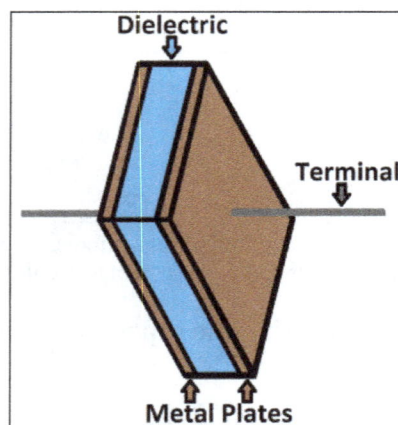

Your standard capacitor sandwich: two metal plates separated by an insulating dielectric.

The dielectric can be made out of all sorts of insulating materials: paper, glass, rubber, ceramic, plastic, or anything that will impede the flow of current.

The plates are made of a conductive material: aluminum, tantalum, silver, or other metals. They're each connected to a terminal wire, which is what eventually connects to the rest of the circuit.

The capacitance of a capacitor - how many farads it has - depends on how it's constructed. More capacitance requires a larger capacitor. Plates with more overlapping surface area provide more capacitance, while more distance between the plates means less capacitance. The material of the dielectric even has an effect on how many farads a cap has. The total capacitance of a capacitor can be calculated with the equation:

$$C = \varepsilon_r \frac{A}{4\pi d}$$

Where ε_r is the dielectric's relative permittivity (a constant value determined by the dielectric material), A is the amount of area the plates overlap each other, and d is the distance between the plates.

Capacitor Working

Electric current is the flow of electric charge, which is what electrical components harness to light up, or spin, or do whatever they do. When current flows into a capacitor, the charges get "stuck" on the plates because they can't get past the insulating dielectric. Electrons - negatively charged particles - are sucked into one of the plates, and it becomes overall negatively charged. The large mass of negative charges on one plate pushes away like charges on the other plate, making it positively charged.

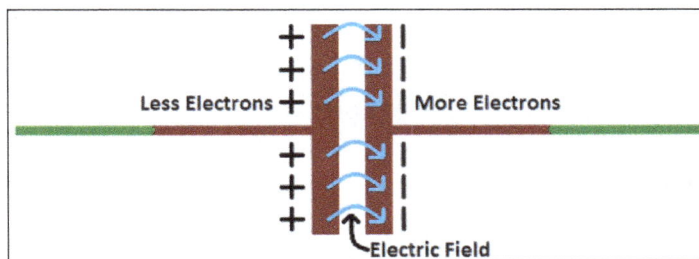

The positive and negative charges on each of these plates attract each other, because that's what opposite charges do. But, with the dielectric sitting between them, as much as they want to come together, the charges will forever be stuck on the plate (until they have somewhere else to go). The stationary charges on these plates create an electric field, which influence electric potential energy and voltage. When charges group together on a capacitor like this, the cap is storing electric energy just as a battery might store chemical energy.

Charging and Discharging

When positive and negative charges coalesce on the capacitor plates, the capacitor becomes charged. A capacitor can retain its electric field - hold its charge - because the positive and negative charges on each of the plates attract each other but never reach each other.

At some point the capacitor plates will be so full of charges that they just can't accept any more. There are enough negative charges on one plate that they can repel any others that try to join. This is where the capacitance (farads) of a capacitor comes into play, which tells you the maximum amount of charge the cap can store.

If a path in the circuit is created, which allows the charges to find another path to each other, they'll leave the capacitor, and it will discharge.

For example, in the circuit below, a battery can be used to induce an electric potential across the capacitor. This will cause equal but opposite charges to build up on each of the plates, until they're so full they repel any more current from flowing. An LED placed in series with the cap could provide a path for the current, and the energy stored in the capacitor could be used to briefly illuminate the LED.

Calculating Charge, Voltage and Current

A capacitor's capacitance - how many farads it has - tells you how much charge it can store. How much charge a capacitor is currently storing depends on the potential difference (voltage) between its plates. This relationship between charge, capacitance, and voltage can be modeled with this equation:

$$Q = CV$$

Charge (Q) stored in a capacitor is the product of its capacitance (C) and the voltage (V) applied to it.

The capacitance of a capacitor should always be a constant, known value. So we can adjust voltage to increase or decrease the cap's charge. More voltage means more charge, less voltage...less charge.

That equation also gives us a good way to define the value of one farad. One farad (F) is the capacity to store one unit of energy (coulombs) per every one volt.

Calculating Current

We can take the charge/voltage/capacitance equation a step further to find out how capacitance and voltage affect current, because current is the rate of flow of charge. The gist of a capacitor's relationship to voltage and current is this: the amount of current through a capacitor depends on both the capacitance and how quickly the voltage is rising or falling. If the voltage across a capacitor swiftly rises, a large positive current will be induced through the capacitor. A slower rise in voltage across a capacitor equates to a smaller current through it. If the voltage across a capacitor is steady and unchanging, no current will go through it.

The equation for calculating current through a capacitor is:

$$i = C\frac{dv}{dt}$$

The dV/dt part of that equation is a derivative (a fancy way of saying instantaneous rate) of voltage over time, it's equivalent to saying "how fast is voltage going up or down at this very moment". The big takeaway from this equation is that if voltage is steady, the derivative is zero, which means current is also zero. This is why current cannot flow through a capacitor holding a steady, DC voltage.

Types of Capacitors

There are all sorts of capacitor types out there, each with certain features and drawbacks which make it better for some applications than others.

When deciding on capacitor types there are a handful of factors to consider:

- Size; Size both in terms of physical volume and capacitance. It's not uncommon for a capacitor to be the largest component in a circuit. They can also be very tiny. More capacitance typically requires a larger capacitor.

- Maximum voltage: Each capacitor is rated for a maximum voltage that can be dropped across it. Some capacitors might be rated for 1.5V, others might be rated for 100V. Exceeding the maximum voltage will usually result in destroying the capacitor.

- Leakage current: Capacitors aren't perfect. Every cap is prone to leaking some tiny amount of current through the dielectric, from one terminal to the other. This tiny current loss (usually nanoamps or less) is called leakage. Leakage causes energy stored in the capacitor to slowly, but surely drain away.

- Equivalent series resistance (ESR): The terminals of a capacitor aren't 100% conductive, they'll always have a tiny amount of resistance (usually less than 0.01Ω) to them. This resistance becomes a problem when a lot of current runs through the cap, producing heat and power loss.

- Tolerance: Capacitors also can't be made to have an exact, precise capacitance. Each cap will be rated for their nominal capacitance, but, depending on the type, the exact value might vary anywhere from ±1% to ±20% of the desired value.

Ceramic Capacitors

The most commonly used and produced capacitor out there is the ceramic capacitor. The name comes from the material from which their dielectric is made.

Ceramic capacitors are usually both physically and capacitance-wise small. It's hard to find a ceramic capacitor much larger than 10µF. A surface-mount ceramic cap is commonly found in a tiny 0402 (0.4mm x 0.2mm), 0603 (0.6mm x 0.3mm) or 0805 package. Through-hole ceramic caps usually look like small (commonly yellow or red) bulbs, with two protruding terminals.

Compared to the equally popular electrolytic caps, ceramics are a more near-ideal capacitor (much lower ESR and leakage currents), but their small capacitance can be limiting. They are usually the least expensive option too. These caps are well-suited for high-frequency coupling and decoupling applications.

Two caps in a through-hole, radial package; a 22pF cap on the left, and a 0.1µF on the right. In the middle, a tiny 0.1µF 0603 surface-mount cap.

Aluminum and Tantalum Electrolytic

Electrolytics are great because they can pack a lot of capacitance into a relatively small volume. If you need a capacitor in the range of 1µF-1mF, you're most likely to find it in an electrolytic form. They're especially well-suited to high-voltage applications because of their relatively high maximum voltage ratings.

Aluminum electrolytic capacitors, the most popular of the electrolytic family, usually look like little tin cans, with both leads extending from the bottom.

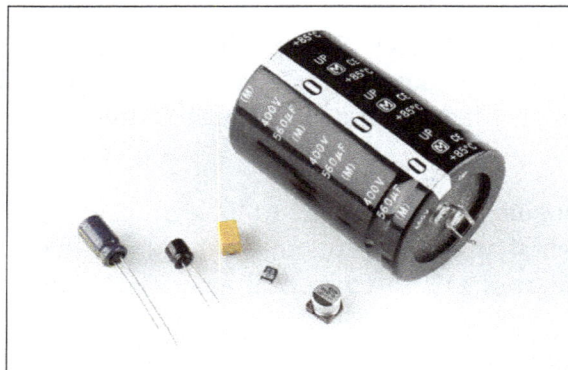

An assortment of through-hole and surface-mount electrolytic capacitors. Notice each has some method for marking the cathode (negative lead).

Unfortunately, electrolytic caps are usually polarized. They have a positive pin - the anode - and a negative pin called the cathode. When voltage is applied to an electrolytic cap, the anode must be at a higher voltage than the cathode. The cathode of an electrolytic capacitor is usually identified with a '-' marking, and a colored strip on the case. The leg of the anode might also be slightly longer as another indication. If voltage is applied in reverse on an electrolytic cap, they'll fail spectacularly (making a pop and bursting open), and permanently. After popping an electrolytic will behave like a short circuit.

These caps also notorious for leakage - allowing small amounts of current (on the order of nA) to run through the dielectric from one terminal to the other. This makes electrolytic caps less-than-ideal for energy storage, which is unfortunate given their high capacity and voltage rating.

Supercapacitors

If you're looking for a capacitor made to store energy, look no further than supercapacitors. These caps are uniquely designed to have very high capacitances, in the range of farads.

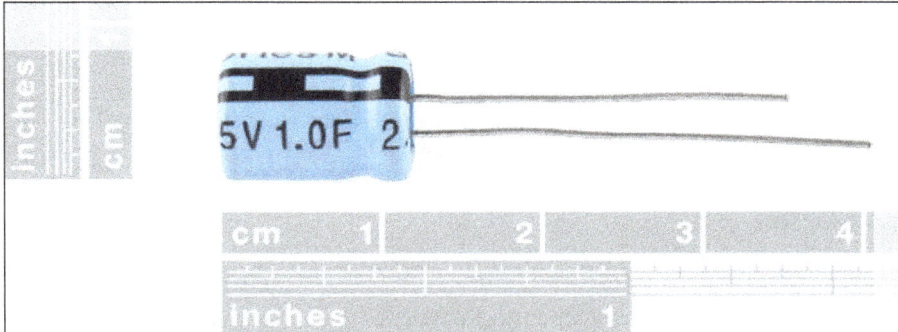

A 1F (!) supercapacitor. High capacitance, but only rated for 2.5V.

While they can store a huge amount of charge, supercaps can't deal with very high voltages. This 10F supercap is only rated for 2.5V max. Any more than that will destroy it. Super caps are commonly placed in series to achieve a higher voltage rating (while reducing total capacitance).

The main application for supercapacitors is in storing and releasing energy, like batteries, which are their main competition. While supercaps can't hold as much energy as an equally sized battery, they can release it much faster, and they usually have a much longer lifespan.

Electrolytic and ceramic caps cover about 80% of the capacitor types out there (and supercaps only about 2%, but they're super!). Another common capacitor type is the film capacitor, which features very low parasitic losses (ESR), making those great for dealing with very high currents.

There are plenty of other less common capacitors. Variable capacitors can produce a range of capacitances, which makes them a good alternative to variable resistors in tuning circuits. Twisted wires or PCBs can create capacitance (sometimes undesired) because each consists of two conductors separated by an insulator. Leyden Jars - a glass jar filled with and surrounded by conductors - are the O.G. of the capacitor family. Finally, of course, flux capacitors (a strange combination of inductor and capacitor) are critical if you ever plan on traveling back to the glory days.

Capacitors in Series/Parallel

Much like resistors, multiple capacitors can be combined in series or parallel to create a combined equivalent capacitance. Capacitors, however, add together in a way that's completely the opposite of resistors.

Capacitors in Parallel

When capacitors are placed in parallel with one another the total capacitance is simply the sum of all capacitances. This is analogous to the way resistors add when in series.

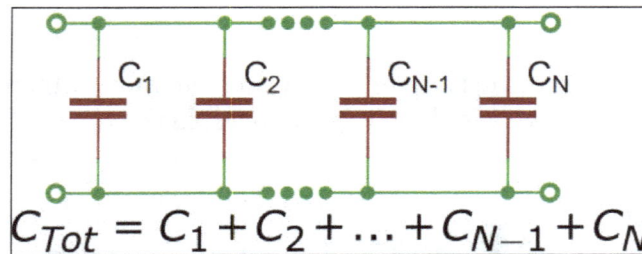

$$C_{Tot} = C_1 + C_2 + \ldots + C_{N-1} + C_N$$

So, for example, if you had three capacitors of values 10μF, 1μF, and 0.1μF in
parallel, the total capacitance would be 11.1μF (10+1+0.1).

Capacitors in Series

Much like resistors are a pain to add in parallel, capacitors get funky when placed in series. The
total capacitance of N capacitors in series is the inverse of the sum of all inverse capacitances.

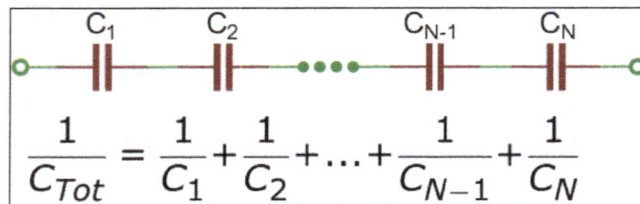

$$\frac{1}{C_{Tot}} = \frac{1}{C_1} + \frac{1}{C_2} + \ldots + \frac{1}{C_{N-1}} + \frac{1}{C_N}$$

If you only have two capacitors in series, you can use the "product-over-sum" method to calculate
the total capacitance:

$$C_{Tot} = \frac{C_1 + C_2}{C_1 + C_2}$$

Taking that equation even further, if you have two equal-valued capacitors in series, the total capacitance is half of their value. For example two 10F supercapacitors in series will produce a total
capacitance of 5F (it'll also have the benefit of doubling the voltage rating of the total capacitor,
from 2.5V to 5V).

Application Examples

There are tons of applications for this nifty little (actually they're usually pretty large) passive component. To give you an idea of their wide range of uses, here are a few examples:

Decoupling (Bypass) Capacitors

A lot of the capacitors you see in circuits, especially those featuring an integrated circuit, are decoupling. A decoupling capacitor's job is to supress high-frequency noise in power supply signals.
They take tiny voltage ripples, which could otherwise be harmful to delicate ICs, out of the voltage
supply.

In a way, decoupling capacitors act as a very small, local power supply for ICs (almost like an un-
interruptible power supply is to computers). If the power supply very temporarily drops its voltage
(which is actually pretty common, especially when the circuit it's powering is constantly switching

its load requirements), a decoupling capacitor can briefly supply power at the correct voltage. This is why these capacitors are also called bypass caps; they can temporarily act as a power source, bypassing the power supply.

Decoupling capacitors connect between the power source (5V, 3.3V, etc.) and ground. It's not uncommon to use two or more different-valued, even different types of capacitors to bypass the power supply, because some capacitor values will be better than others at filtering out certain frequencies of noise.

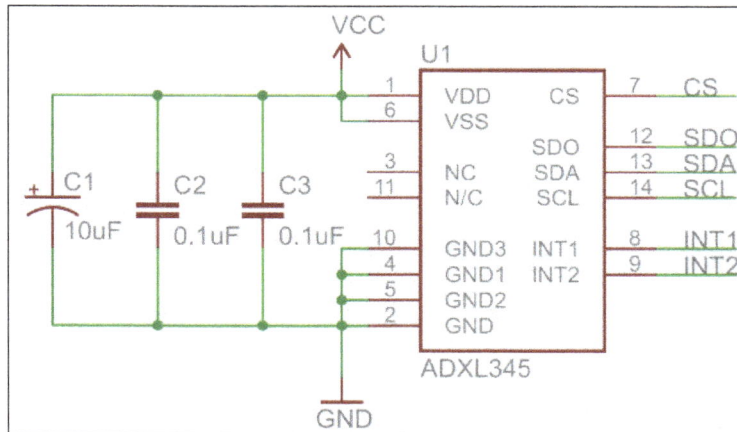

In this schematic, three decoupling capacitors are used to help reduce the noise in an accelerometer's voltage supply. Two ceramic 0.1µF and one tantalum electrolytic 10µF split decoupling duties.

While it seems like this might create a short from power to ground, only high-frequency signals can run through the capacitor to ground. The DC signal will go to the IC, just as desired. Another reason these are called bypass capacitors is because the high frequencies (in the kHz-MHz range) bypass the IC, instead running through the capacitor to get to ground.

When physically placing decoupling capacitors, they should always be located as close as possible to an IC. The further away they are, they less effective they'll be.

Here's the physical circuit layout from the schematic above. The tiny, black IC is surrounded by two 0.1µF capacitors (the brown caps) and one 10µF electrolytic tantalum capacitor (the tall, black/grey rectangular cap).

To follow good engineering practice, always add at least one decoupling capacitor to every IC. Usually 0.1µF is a good choice, or even add some 1µF or 10µF caps. They're a cheap addition, and they help make sure the chip isn't subjected to big dips or spikes in voltage.

Power Supply Filtering

Diode rectifiers can be used to turn the AC voltage coming out of your wall into the DC voltage required by most electronics. But diodes alone can't turn an AC signal into a clean DC signal, they need the help of capacitors! By adding a parallel capacitor to a bridge rectifier, rectified signals like this:

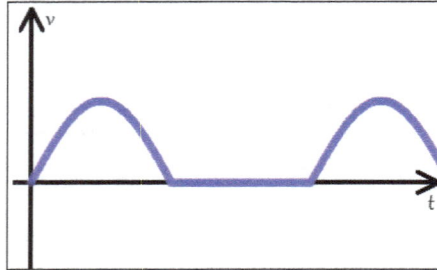

Can be turned into a near-level DC signal like this:

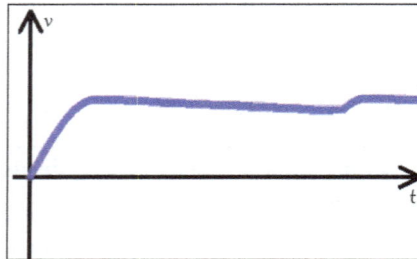

Capacitors are stubborn components; they'll always try to resist sudden changes in voltage. The filter capacitor will charge up as the rectified voltage increases. When the rectified voltage coming into the cap starts its rapid decline, the capacitor will access its bank of stored energy, and it'll discharge very slowly, supplying energy to the load. The capacitor shouldn't fully discharge before the input rectified signal starts to increase again, recharging the cap. This dance plays out many times a second, over-and-over as long as the power supply is in use.

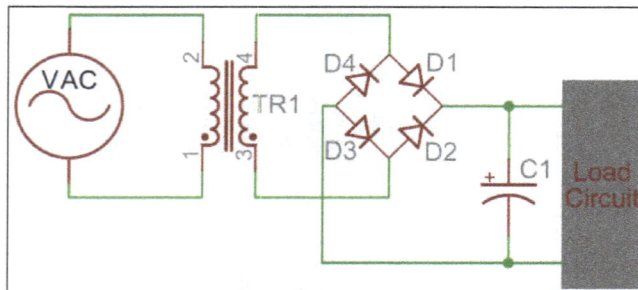

An AC-to-DC power supply circuit. The filter cap (C1) is critical in smoothing out the DC signal sent to the load circuit.

If you tear apart any AC-to-DC power supply, you're bound to find at least one rather large capacitor.

Energy Storage and Supply

It seems obvious that if a capacitor stores energy, one of its many applications would be supplying that energy to a circuit, just like a battery. The problem is capacitors have a much lower energy density than batteries; they just can't pack as much energy as an equally sized chemical battery (but that gap is narrowing!).

The upside of capacitors is they usually lead longer lives than batteries, which makes them a better choice environmentally. They're also capable of delivering energy much faster than a battery, which makes them good for applications which need a short, but high burst of power. A camera flash might get its power from a capacitor (which, in turn, was probably charged by a battery).

Battery or Capacitor		
	Battery	Capacitor
Capacity	✓	
Energy Density	✓	
Charge/Discharge Rate		✓
Life Span		✓

Signal Filtering

Capacitors have a unique response to signals of varying frequencies. They can block out low-frequency or DC signal-components while allowing higher frequencies to pass right through. They're like a bouncer at a very exclusive club for high frequencies only.

Filtering signals can be useful in all sorts of signal processing applications. Radio receivers might use a capacitor (among other components) to tune out undesired frequencies.

Another example of capacitor signal filtering is passive crossover circuits inside speakers, which separate a single audio signal into many. A series capacitor will block out low frequencies, so the remaining high-frequency parts of the signal can go to the speaker's tweeter. In the low-frequency passing, subwoofer circuit, high-frequencies can mostly be shunted to ground through the parallel capacitor.

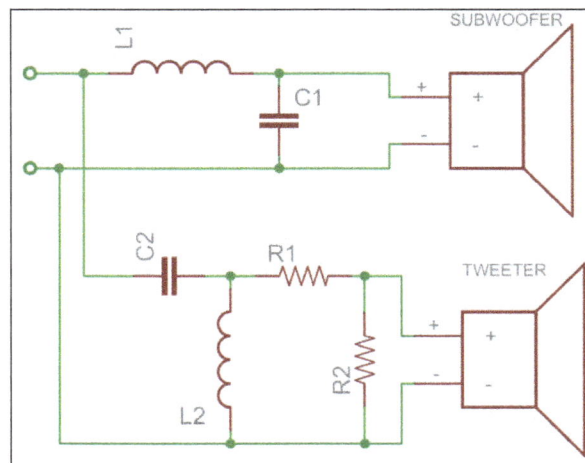

A very simple example of an audio crossover circuit. The capacitor will block out low frequencies, while the inductor blocks out high frequencies. Each can be used to deliver the proper signal to tuned audio drivers.

Inductor

The inductor is a passive component which stores the electrical energy in the magnetic field when the electric current passes through it. Or we can say that the inductor is an electrical device which possesses the inductance.

The inductor is made of wire which has the property of inductance, i.e., opposes the flow of current. The inductance of wire increases by increasing the number of turns. The alphabet 'L' is used for representing the inductor, and it is measured in Henry. The inductance characterises the inductor. The figure below shows the symbolic representation of inductor.

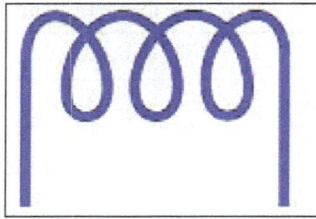

The electric current I flows through the coil generates the magnetic field around it. Consider the magnetic field generates the flux φ when current flows through it. The ratio of the flux and the current gives inductances:

$$L = \frac{\phi}{I}$$

The inductance of the circuit depends on the current paths and the magnetic permeability of the nearer material. The magnetic permeability shows the ability of the material to forms the magnetic field.

Types of Inductor

The inductors are classified into two types.

1. Air Cored Inductor (wound on non-ferrite material): The inductor in which either the core is completely absent or ceramic material is used for making the core such type of inductor is known as the air-cored inductor.

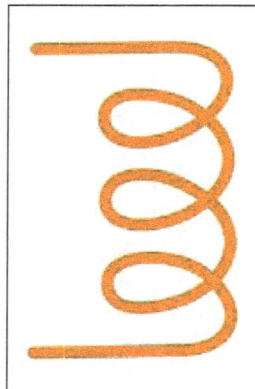

The ceramic material has the very low thermal coefficient of expansion. The low thermal coefficient of expansion means the shape of material remains same even with the rise of temperature. The ceramic material has no magnetic properties. The permeability of the inductor remains same due to the ceramic material.

In air core-inductor, the only work of the core is to give the coil a particular shape. The air cored structure has many advantages like they reduce the core losses and increases the quality factor. The air cored inductor is used for high-frequency applications work where low inductance is required.

2. Iron Core Inductor (wound on ferrite core): It is a fixed value inductor in which the iron core is kept between the coil. The iron-cored inductor is used in the filter circuit for smoothing out the ripple voltage, it is also used as a choke in fluorescent tube light, in industrial power supplies and inverter system etc.

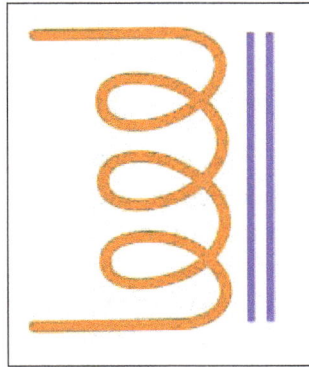

Working of Inductor

The inductor is an electrical device used for storing the electrical energy in the form of the magnetic field. It is constructed by wounding the wire on the core. The cores are made of ceramic material, iron or by the air. The core may be toroidal or E- shaped.

The coil-carrying the electric current induces the magnetic field around the conductor. The intensity of the magnetic field increases if the core is placed between the coil. The core provides the low reluctance path to the magnetic flux.

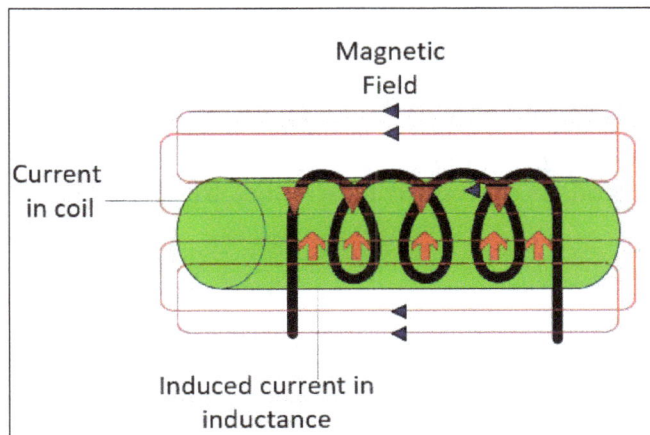

The magnetic field induces the EMF in the coil which causes the current. And according to Lenz's law, the causes always oppose the effect. Here, the current is the causes, and it is induced because of the voltage. Thus, the EMF oppose the change of current that changes the magnetic field. The current which reduces because of the inductance is known as the inductive reactance. The inductive reactance increases with the increase of the number of turn of coils.

Active Components

In an electrical system, active components are those that require electrical power to operate. This could include the transistors, diodes and other integrated circuits but would exclude system components such as the chassis, capacitors or enclosures that do not require electrical power to operate.

Diode

A diode is defined as a two-terminal electronic component that only conducts current in one direction (so long as it is operated within a specified voltage level). An ideal diode will have zero resistance in one direction, and infinite resistance in the reverse direction. Although in the real world, diode's cannot achieve zero or infinite resistance. Instead, a diode will have negligible resistance in one direction (to allow current flow), and a very high resistance in the reverse direction (to prevent current flow). A diode is effectively like a valve for an electrical circuit.

Semiconductor diodes are the most common type of diode. These diodes begin conducting electricity only if a certain threshold voltage is present in the forward direction (i.e. the "low resistance" direction). The diode is said to be "forward biased" when conducting current in this direction. When connected within a circuit in the reverse direction (i.e. the "high resistance" direction), the diode is said to be "reverse biased".

A diode only blocks current in the reverse direction (i.e. when it is reversing biased) while the reverse voltage is within a specified range. Above this range, the reverse barrier breaks. The voltage at which this breakdown occurs is called the "reverse breakdown voltage". When the voltage of the circuit is higher than the reverse breakdown voltage, the diode is able to conduct electricity in the reverse direction (i.e. the "high resistance" direction). This is why in practice we say diode's have a high resistance in the reverse direction – not an infinite resistance.

A PN junction is the simplest form of the semiconductor diode. In ideal conditions, this PN junction behaves as a short circuit when it is forward biased and as an open circuit when it is in the reverse biased. The name diode is derived from "di–ode" which means a device that has two electrodes.

Diode Symbol

The symbol of a diode is shown below. The arrowhead points in the direction of conventional current flow in the forward biased condition. That means the anode is connected to the p side and cathode is connected to the n side.

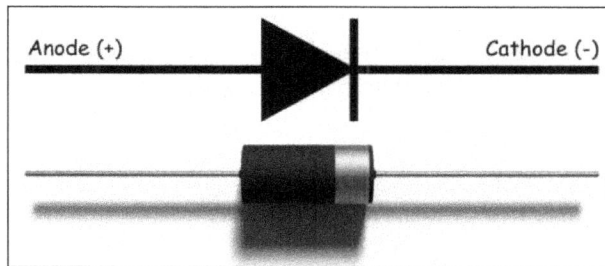

We can create a simple PN junction diode by doping pentavalent or donor impurity in one portion and trivalent or acceptor impurity in other portion of silicon or germanium crystal block. These dopings make a PN junction at the middle part of the block. We can also form a PN junction by joining a p-type and n-type semiconductor together with a special fabrication technique. The terminal connected to the p-type is the anode. The terminal connected to the n-type side is the cathode.

Working Principle of Diode

A diode's working principle depends on the interaction of n-type and p-type semiconductors. An n-type semiconductor has plenty of free electrons and a very few numbers of holes. In other words, we can say that the concentration of free electrons is high and that of holes is very low in an n-type semiconductor. Free electrons in the n-type semiconductor are referred as majority charge carriers, and holes in the n-type semiconductor are referred to as minority charge carriers.

A p-type semiconductor has a high concentration of holes and low concentration of free electrons. Holes in the p-type semiconductor are majority charge carriers, and free electrons in the p-type semiconductor are minority charge carriers.

Unbiased Diode

Now let us see what happens when one n-type region and one p-type region come in contact. Here due to concentration difference, majority carriers diffuse from one side to another. As the concentration of holes is high in the p-type region and it is low in the n-type region, the holes start diffusing from the p-type region to n-type region. Again the concentration of free electrons is high in the n-type region and it is low in the p-type region and due to this reason, free electrons start diffusing from the n-type region to the p-type region.

The free electrons diffusing into the p-type region from the n-type region would recombine with holes available there and create uncovered negative ions in the p-type region. In the same way, the holes diffusing into the n-type region from the p-type region would recombine with free electrons available there and create uncovered positive ions in the n-type region.

In this way, there would a layer of negative ions in the p-type side and a layer of positive ions in the n-type region appear along the junction line of these two types of semiconductor. The layers of uncovered positive ions and uncovered negative ions form a region at the middle of the diode where no charge carrier exists since all the charge carriers get recombined here in this region. Due to lack of charge carriers, this region is called depletion region.

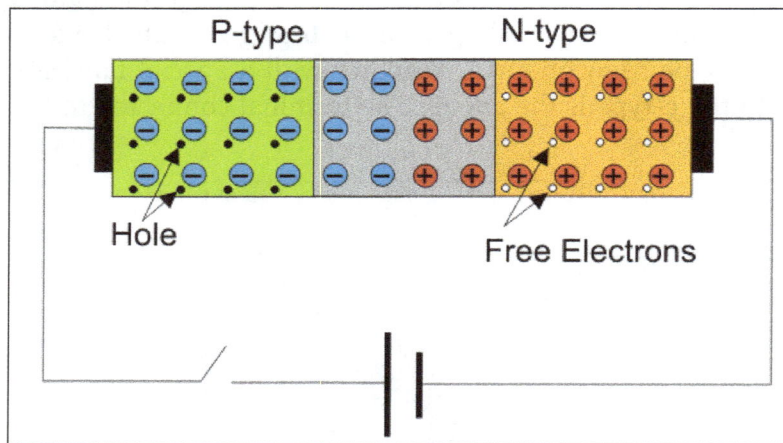

After the formation of the depletion region, there is no more diffusion of charge carriers from one side to another in the diode. This is because due to the electric field appeared across the depletion region will prevent further migration of charge carriers from one side to another. The potential of the layer of uncovered positive ions in the n-type side would repeal the holes in the p-type side and the potential of the layer of uncovered negative ions in the p-type side would repeal the free electrons in the n-type side. That means a potential barrier is created across the junction to prevent further diffusion of charge carriers.

Forward Biased Diode

Now let us see what happens if positive terminal of a source is connected to the p-type side and the negative terminal of the source is connected to the n-type side of the diode and if we increase the voltage of this source slowly from zero.

In the beginning, there is no current flowing through the diode. This is because although there is an external electrical field applied across the diode but still the majority charge carriers do not get sufficient influence of the external field to cross the depletion region. As we told that the depletion region acts as a potential barrier against the majority charge carriers. This potential barrier is called forward potential barrier. The majority charge carriers start crossing the forward potential barrier only when the value of externally applied voltage across the junction is more than the potential of the forward barrier. For silicon diodes, the forward barrier potential is 0.7 volt and for germanium diodes, it is 0.3 volt. When the externally applied forward voltage across the diode becomes more than the forward barrier potential, the free majority charge carriers start crossing the barrier and contribute the forward diode current. In that situation, the diode would behave as a short-circuited path and the forward current gets limited by only externally connected resistors to the diode.

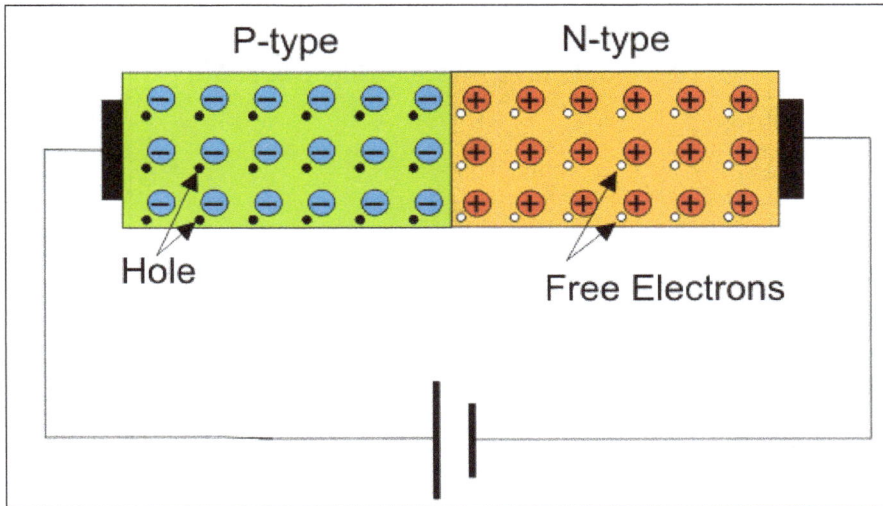

Reverse Biased Diode

Now let us see what happens if we connect negative terminal of the voltage source to the p-type side and positive terminal of the voltage source to the n-type side of the diode. At that condition, due to electrostatic attraction of negative potential of the source, the holes in the p-type region would be shifted more away from the junction leaving more uncovered negative ions at the junction. In the same way, the free electrons in the n-type region would be shifted more away from the junction towards the positive terminal of the voltage source leaving more uncovered positive ions in the junction. As a result of this phenomenon, the depletion region becomes wider. This condition of a diode is called the reverse biased condition. At that condition, no majority carriers cross the junction as they go away from the junction. In this way, a diode blocks the flow of current when it is reverse biased.

There are always some free electrons in the p-type semiconductor and some holes in the n-type semiconductor. These opposite charge carriers in a semiconductor are called minority charge carriers. In the reverse biased condition, the holes find themselves in the n-type side would easily cross the reverse biased depletion region as the field across the depletion region does not present rather it helps minority charge carriers to cross the depletion region. As a result, there is a tiny current flowing through the diode from positive to the negative side. The amplitude of this current is very small as the number of minority charge carriers in the diode is very small. This current is called reverse saturation current.

If the reverse voltage across a diode gets increased beyond a safe value, due to higher electrostatic force and due to a higher kinetic energy of minority charge carriers colliding with atoms, a number of covalent bonds get broken to contribute a huge number of free electron-hole pairs in the diode and the process is cumulative. The huge number of such generated charge carriers would contribute a huge reverse current in the diode. If this current is not limited by an external resistance connected to the diode circuit, the diode may permanently be destroyed.

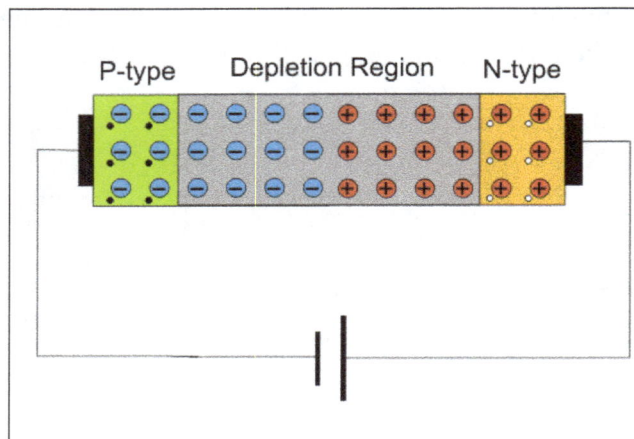

Types of Diode

The types of diode include:

- Zener diode

- P-N junction diode

- Tunnel diode

- Varactor diode

- Schottky diode

- Photodiode

- PIN diode

- Laser diode

- Avalanche diode

- Light emitting diode.

Transistor

Transistor is a semiconductor device that can both conduct and insulate. A transistor can act as a switch and an amplifier. It converts audio waves into electronic waves and resistor, controlling electronic current. Transistors have very long life, smaller in size, can operate on lower voltage supplies for greater safety and required no filament current. The first transistor was fabricated with germanium. A transistor performs the same function as a vacuum tube triode, but using semiconductor junctions instead of heated electrodes in a vacuum chamber. It is the fundamental building block of modern electronic devices and found everywhere in modern electronic systems.

A transistor is a three terminal device. Namely,

- Base: This is responsible for activating the transistor.

- Collector: This is the positive lead.

- Emitter: This is the negative lead.

The basic idea behind a transistor is that it lets you control the flow of current through one channel by varying the intensity of a much smaller current that's flowing through a second channel.

Types of Transistors

There are two types of transistors in present; they are bipolar junction transistor (BJT), field effect transistors (FET). A small current is flowing between the base and the emitter; base terminal can control a larger current flow between the collector and the emitter terminals. For a field-effect transistor, it also has the three terminals, they are gate, source, and drain, and a voltage at the gate can control a current between source and drain. The simple diagrams of BJT and FET are shown in figure below:

Bipolar Junction Transistor(BJT).

Field Effect Transistors(FET).

As you can see, transistors come in a variety of different sizes and shapes. One thing all of these transistors have in common is that they each have three leads.

Bipolar Junction Transistor

A Bipolar Junction Transistor (BJT) has three terminals connected to three doped semiconductor regions. It comes with two types, P-N-P and N-P-N.

P-N-P transistor, consisting of a layer of N-doped semiconductor between two layers of P-doped material. The base current entering in the collector is amplified at its output.

That is when PNP transistor is ON when its base is pulled low relative to the emitter. The arrows of PNP transistor symbol the direction of current flow when the device is in forward active mode.

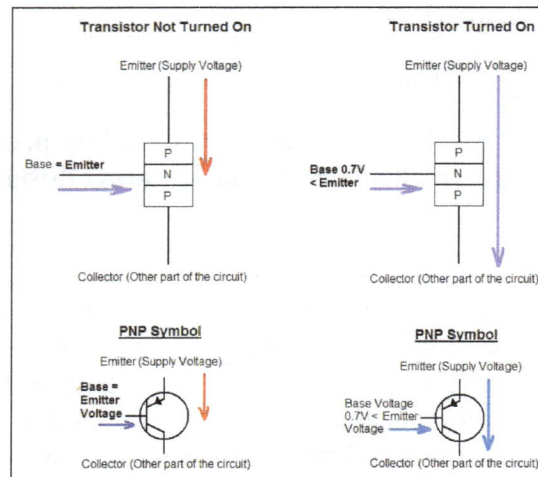

N-P-N transistor consisting a layer of P-doped semiconductor between two layers of N-doped material. By amplifying current the base we get the high collector and emitter current.

That is when NPN transistor is ON when its base is pulled low relative to the emitter. When the transistor is in ON state, current flow is in between the collector and emitter of the transistor. Based on minority carriers in P-type region the electrons moving from emitter to collector. It allows the greater current and faster operation; because of this reason most bipolar transistors used today are NPN.

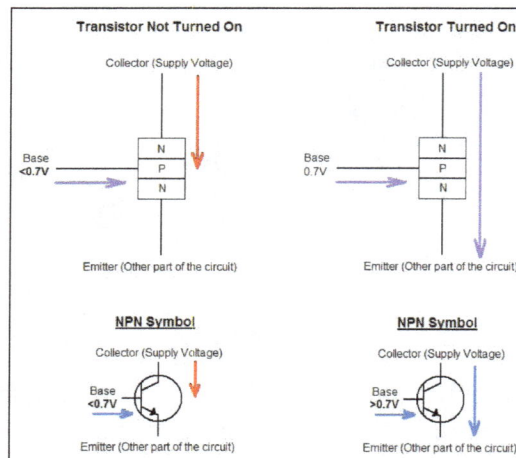

Field Effect Transistor (FET)

The field effect transistor is a unipolar transistor, N-channel FET or P-channel FET are used for conduction. The three terminals of FET are source, gate and drain. The basic n-channel and p-channel FET's are shown above. For an n-channel FET, the device is constructed from n-type material. Between the source and drain then-type material acts as a resistor.

This transistor controls the positive and negative carriers with respect to holes or electrons. FET channel is formed by moving of positive and negative charge carriers. The channel of FET which is made by silicon.

There are many types of FET's, MOSFET, JFET and etc. The applications of FET's are in low noise amplifier, buffer amplifier and analog switch.

Bipolar Junction Transistor Biasing

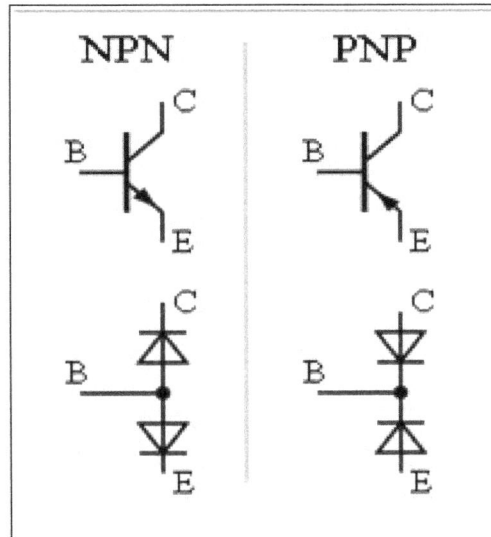

Transistors are the most important semiconductor active devices essential for almost all circuits. They are used as electronic switches, amplifiers etc in circuits. Transistors may be NPN, PNP, FET, JFET etc which have different functions in electronic circuits. For the proper working of the circuit, it is necessary to bias the transistor using resistor networks. Operating point is the point on the output characteristics that shows the Collector-Emitter voltage and the Collector current with no input signal. The Operating point is also known as the Bias point or Q-Point (Quiescent point).

Biasing is referred to provide resistors, capacitors or supply voltage etc to provide proper operating characteristics of the transistors. DC biasing is used to obtain DC collector current at a particular collector voltage. The value of this voltage and current are expressed in terms of the Q-Point. In a transistor amplifier configuration, the IC (max) is the maximum current that can flow through the transistor and VCE (max) is the maximum voltage applied across the device. To work the transistor as an amplifier, a load resistor RC must be connected to the collector. Biasing set the DC operating voltage and current to the correct level so that the AC input signal can be properly amplified by the transistor. The correct biasing point is somewhere between the fully ON or fully OFF states of the transistor. This central point is the Q-Point and if the transistor is properly biased, the Q-point will be the central operating point of the transistor. This helps the output current to increase and decrease as the input signal swings through the complete cycle.

For setting the correct Q-Point of the transistor, a collector resistor is used to set the collector current to a constant and steady value without any signal in its base. This steady DC operating point is set by the value of the supply voltage and the value of the base biasing resistor. Base bias resistors are used in all the three transistor configurations like common base, common collector and Common emitter configurations.

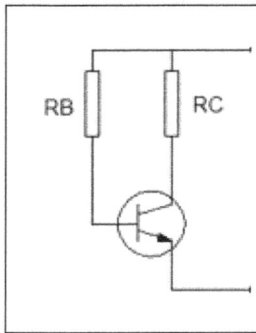

Current Biasing   Feedback Biasing   Double Feedback Biasing

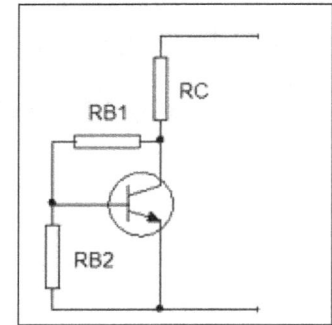

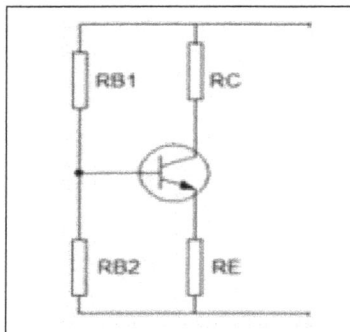

Voltage Divider Biasing   Double Base Biasing

Modes of Biasing

Following are the different modes of transistor base biasing:

1. Current biasing:

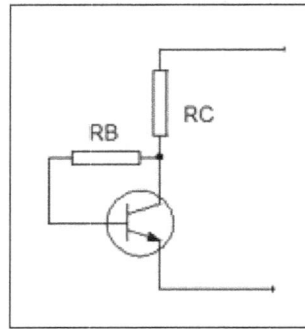

As shown in the Figure, two resistors RC and RB are used to set the base bias. These resistors establish the initial operating region of the transistor with a fixed current bias.

The transistor forward biases with a positive base bias voltage through RB. The forward base-Emitter voltage drop is 0.7 volts. Therefore the current through RB is $I_B = (V_{cc} - V_{BE}) / I_B$.

2. Feedback biasing:

Figure shows the transistor biasing by the use of a feedback resistor. The base bias is obtained from the collector voltage. The collector feedback ensures that the transistor is always biased in the active region. When the collector current increases, the voltage at the collector drops. This reduces the base drive which in turn reduces the collector current. This feedback configuration is ideal for transistor amplifier designs.

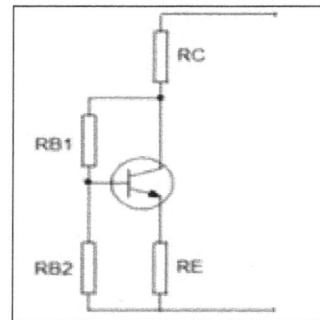

3. Double Feedback Biasing:

Figure above shows how the biasing is achieved using double feedback resistors.

By using two resistors RB1 and RB2 increases the stability with respect to the variations in Beta by increasing the current flow through the base bias resistors. In this configuration, the current in RB1 is equal to 10 % of the collector current.

4. Voltage Dividing Biasing:

The Voltage divider biasing in which two resistors RB1 and RB2 are connected to the base of the transistor forming a voltage divider network. The transistor gets biases by the voltage drop across RB2. This kind of biasing configuration is used widely in amplifier circuits.

5. Double Base Biasing:

A double feedback for stabilization. It uses both Emitter and Collector base feedback to improve the stabilization through controlling the collector current. Resistor values should be selected so as to set the voltage drop across the Emitter resistor 10% of the supply voltage and the current through RB1, 10% of the collector current.

Advantages of Transistor

1. Smaller mechanical sensitivity.

2. Lower cost and smaller in size, especially in small-signal circuits.

3. Low operating voltages for greater safety, lower costs and tighter clearances.

4. Extremely long life.

5. No power consumption by a cathode heater.

6. Fast switching.

Integrated Circuit

Integrated circuit (IC), also called microelectronic circuit, microchip, or chip is an assembly of electronic components, fabricated as a single unit, in which miniaturized active devices (e.g., transistors and diodes) and passive devices (e.g., capacitors and resistors) and their interconnections are built up on a thin substrate of semiconductor material (typically silicon). The resulting circuit is thus a small monolithic "chip," which may be as small as a few square centimetres or only a few square millimetres. The individual circuit components are generally microscopic in size.

Integrated circuits have their origin in the invention of the transistor in 1947 by William B. Shockley and his team at the American Telephone and Telegraph Company's Bell Laboratories. Shockley's team (including John Bardeen and Walter H. Brattain) found that, under the right circumstances, electrons would form a barrier at the surface of certain crystals, and they learned to control the flow of electricity through the crystal by manipulating this barrier. Controlling electron flow through a crystal allowed the team to create a device that could perform certain electrical operations, such as signal amplification, that were previously done by vacuum tubes. They named this device a transistor, from a combination of the words transfer and resistor. The study of methods of creating electronic devices using solid materials became known as solid-state electronics. Solid-state devices proved to be much sturdier, easier to work with, more reliable, much smaller, and less expensive than vacuum tubes. Using the same principles and materials, engineers soon learned to create other electrical components, such as resistors and capacitors. Now that electrical devices could be made so small, the largest part of a circuit was the awkward wiring between the devices.

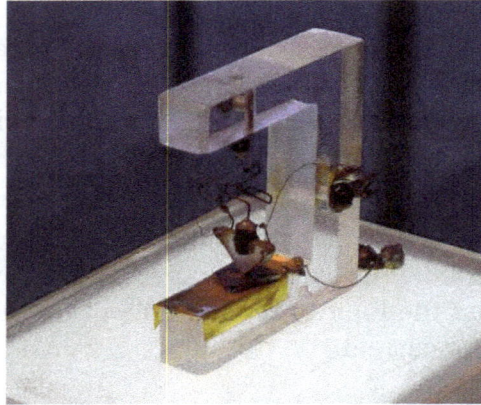

The first transistor, invented by American physicists John Bardeen, Walter H. Brattain, and William B. Shockley.

In 1958 Jack Kilby of Texas Instruments, Inc., and Robert Noyce of Fairchild Semiconductor Corporation independently thought of a way to reduce circuit size further. They laid very thin paths of metal (usually aluminum or copper) directly on the same piece of material as their devices. These small paths acted as wires. With this technique an entire circuit could be "integrated" on a single piece of solid material and an integrated circuit (IC) thus created. ICs can contain hundreds of thousands of individual transistors on a single piece of material the size of a pea. Working with that many vacuum tubes would have been unrealistically awkward and expensive. The invention of the integrated circuit made technologies of the Information Age feasible. ICs are now used extensively in all walks of life, from cars to toasters to amusement park rides.

Basic IC Types

Analog versus Digital Circuits

Analog, or linear, circuits typically use only a few components and are thus some of the simplest types of ICs. Generally, analog circuits are connected to devices that collect signals from the environment or send signals back to the environment. For example, a microphone converts fluctuating vocal sounds into an electrical signal of varying voltage. An analog circuit then modifies the signal in some useful way—such as amplifying it or filtering it of undesirable noise. Such a signal might then be fed back to a loudspeaker, which would reproduce the tones originally picked up by the microphone. Another typical use for an analog circuit is to control some device in response to continual changes in the environment. For example, a temperature sensor sends a varying signal to a thermostat, which can be programmed to turn an air conditioner, heater, or oven on and off once the signal has reached a certain value.

A digital circuit, on the other hand, is designed to accept only voltages of specific given values. A circuit that uses only two states is known as a binary circuit. Circuit design with binary quantities, "on" and "off" representing 1 and 0 (i.e., true and false), uses the logic of Boolean algebra. (Arithmetic is also performed in the binary number system employing Boolean algebra.) These basic elements are combined in the design of ICs for digital computers and associated devices to perform the desired functions.

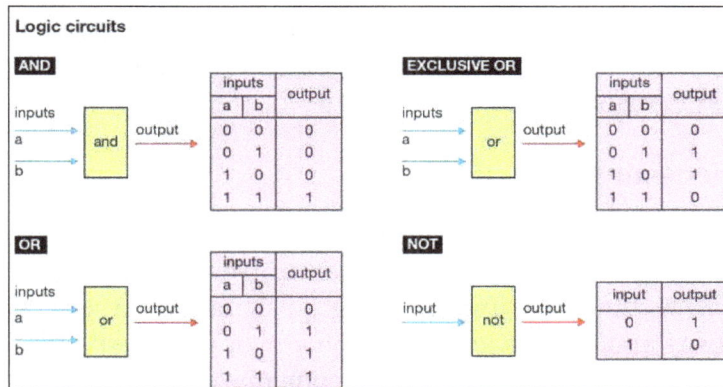

Logic circuits

	inputs a	inputs b	output
AND	0	0	0
	0	1	0
	1	0	0
	1	1	1

inputs a, b → and → output

	inputs a	inputs b	output
EXCLUSIVE OR	0	0	0
	0	1	1
	1	0	1
	1	1	0

inputs a, b → or → output

	inputs a	inputs b	output
OR	0	0	0
	0	1	1
	1	0	1
	1	1	1

inputs a, b → or → output

	input	output
NOT	0	1
	1	0

input → not → output

Logic circuit: Different combinations of logic circuits.

Microprocessor Circuits

Microprocessors are the most-complicated ICs. They are composed of billions of transistors that have been configured as thousands of individual digital circuits, each of which performs some specific logic function. A microprocessor is built entirely of these logic circuits synchronized to each other. Microprocessors typically contain the central processing unit (CPU) of a computer.

Just like a marching band, the circuits perform their logic function only on direction by the bandmaster. The bandmaster in a microprocessor, so to speak, is called the clock. The clock is a signal that quickly alternates between two logic states. Every time the clock changes state, every logic circuit in the microprocessor does something. Calculations can be made very quickly, depending on the speed (clock frequency) of the microprocessor.

Microprocessors contain some circuits, known as registers that store information. Registers are predetermined memory locations. Each processor has many different types of registers. Permanent registers are used to store the preprogrammed instructions required for various operations (such as addition and multiplication). Temporary registers store numbers that are to be operated on and also the result. Other examples of registers include the program counter (also called the instruction pointer), which contains the address in memory of the next instruction; the stack pointer (also called the stack register), which contains the address of the last instruction put into an area of memory called the stack; and the memory address register, which contains the address of where the data to be worked on is located or where the data that has been processed will be stored.

Microprocessors can perform billions of operations per second on data. In addition to computers, microprocessors are common in video game systems, televisions, cameras, and automobiles.

Memory Circuits

Microprocessors typically have to store more data than can be held in a few registers. This additional information is relocated to special memory circuits. Memory is composed of dense arrays of parallel circuits that use their voltage states to store information. Memory also stores the temporary sequence of instructions, or program, for the microprocessor.

Manufacturers continually strive to reduce the size of memory circuits—to increase capability without increasing space. In addition, smaller components typically use less power, operate more efficiently, and cost less to manufacture.

Digital Signal Processors

A signal is an analog waveform—anything in the environment that can be captured electronically. A digital signal is an analog waveform that has been converted into a series of binary numbers for quick manipulation. As the name implies, a digital signal processor (DSP) processes signals digitally, as patterns of 1s and 0s. For instance, using an analog-to-digital converter, commonly called an A-to-D or A/D converter, a recording of someone's voice can be converted into digital 1s and 0s. The digital representation of the voice can then be modified by a DSP using complex mathematical formulas. For example, the DSP algorithm in the circuit may be configured to recognize gaps between spoken words as background noise and digitally remove ambient noise from the waveform. Finally, the processed signal can be converted back (by a D/A converter) into an analog signal for listening. Digital processing can filter out background noise so fast that there is no discernible delay and the signal appears to be heard in "real time." For instance, such processing enables "live" television broadcasts to focus on a quarterback's signals in an American gridiron football game.

DSPs are also used to produce digital effects on live television. For example, the yellow marker lines displayed during the football game are not really on the field; a DSP adds the lines after the cameras shoot the picture but before it is broadcast. Similarly, some of the advertisements seen on stadium fences and billboards during televised sporting events are not really there.

Application-specific ICs

An application-specific IC (ASIC) can be either a digital or an analog circuit. As their name implies, ASICs are not reconfigurable; they perform only one specific function. For example, a speed controller IC for a remote control car is hard-wired to do one job and could never become a microprocessor. An ASIC does not contain any ability to follow alternate instructions.

Radio-frequency ICs

Radio-frequency ICs (RFICs) are widely used in mobile phones and wireless devices. RFICs are analog circuits that usually run in the frequency range of 3 kHz to 2.4 GHz (3,000 hertz to 2.4 billion hertz), circuits that would work at about 1 THz (1 trillion hertz) being in development. They are usually thought of as ASICs even though some may be configurable for several similar applications.

Most semiconductor circuits that operate above 500 MHz (500 million hertz) cause the electronic components and their connecting paths to interfere with each other in unusual ways. Engineers must use special design techniques to deal with the physics of high-frequency microelectronic interactions.

Monolithic Microwave ICs

A special type of RFIC is known as a monolithic microwave IC (MMIC; also called microwave monolithic IC). These circuits usually run in the 2- to 100-GHz range, or microwave frequencies, and are used in radar systems, in satellite communications, and as power amplifiers for cellular telephones.

Just as sound travels faster through water than through air, electron velocity is different through each type of semiconductor material. Silicon offers too much resistance for microwave-frequency circuits, and so the compound gallium arsenide (GaAs) is often used for MMICs. Unfortunately, GaAs is mechanically much less sound than silicon. It breaks easily, so GaAs wafers are usually much more expensive to build than silicon wafers.

Designing ICs

All ICs use the same basic principles of voltage (V), current (I), and resistance (R). In particular, equations based on Ohm's law, $V = IR$, determine many circuit design choices. Design engineers must also be familiar with the properties of various electronic components needed for different applications.

Analog Design

An analog circuit takes an infinitely variable real-world voltage or current and modifies it in some useful way. The signal might be amplified, compared with another signal, mixed with other signals, separated from other signals, examined for value, or otherwise manipulated. For the design of this type of circuit, the choice of every individual component, size, placement, and connection is crucial. Unique decisions abound—for instance, whether one connection should be slightly wider than another connection, whether one resistor should be oriented parallel or perpendicular to another, or whether one wire can lie over the top of another. Every small detail affects the final performance of the end product.

When integrated circuits were much simpler, component values could be calculated by hand. For instance, a specific amplification value (gain) of an amplifier could typically be calculated from the ratio of two specific resistors. The current in the circuit could then be determined, using the resistor value required for the amplifier gain and the supply voltage used. As designs became more complex, laboratory measurements were used to characterize the devices. Engineers drew graphs of device characteristics across several variables and then referred to those graphs as they needed information for their calculations. As scientists improved their characterization of the intricate physics of each device, they developed complex equations that took into account subtle effects that were not apparent from coarse laboratory measurements. For example, a transistor works very differently at different frequencies, sizes, orientations, and placements. In particular, scientists found parasitic components (unwanted effects, usually resistance and capacitance) that are inherent in the way the devices are built. Parasitics become more problematic as the circuitry becomes more sophisticated and smaller and as it runs at higher frequencies.

Although parasitic components in a circuit can now be accounted for by sophisticated equations, such calculations are very time-consuming to do by hand. For this work computers have become indispensable. In particular, a public-domain circuit-analysis program developed at the University of California, Berkeley, during the 1970s, SPICE (Simulation Program with Integrated Circuit Emphasis), and various proprietary models designed for use with it are ubiquitous in engineering courses and in industry for analog circuit design. SPICE has equations for transistors, capacitors, resistors, and other components, as well as for lengths of wires and for turns in wires, and it can reduce the calculation of circuit interactions to hours from the months formerly required for hand calculations.

Digital Design

Since digital circuits involve millions of times as many components as analog circuits, much of the design work is done by copying and reusing the same circuit functions, especially by using digital design software that contains libraries of prestructured circuit components. The components available in such a library are of similar height, contain contact points in predefined locations, and have other rigid conformities so that they fit together regardless of how the computer configures a layout. While SPICE is perfectly adequate for analyzing analog circuits, with equations that describe individual components, the complexity of digital circuits requires a less-detailed approach. Therefore, digital analysis software ignores individual components for mathematical models of entire preconfigured circuit blocks (or logic functions).

Whether analog or digital circuitry is used depends on the function of a circuit. The design and layout of analog circuits are more demanding of teamwork, time, innovation, and experience, particularly as circuit frequencies get higher, though skilled digital designers and layout engineers can be of great benefit in overseeing an automated process as well. Digital design emphasizes different skills from analog design.

Mixed-signal Design

For designs that contain both analog and digital circuitry (mixed-signal chips), standard analog and digital simulators are not sufficient. Instead, special behavioral simulators are used, employing the same simplifying idea behind digital simulators to model entire circuits rather than individual transistors. Behavioral simulators are designed primarily to speed up simulations of the analog side of a mixed-signal chip.

The difficulty with behavioral simulation is making sure that the model of the analog circuit function is accurate. Since each analog circuit is unique, it seems as though one must design the system twice—once to design the circuitry and once to design the model for the simulator.

Fabricating ICs

Making a Base Wafer

The substrate material, or base wafer, on which ICs are built, is a semiconductor, such as silicon or gallium arsenide. In order to obtain consistent performance, the semiconductor must be extremely pure and a single crystal. The basic technique for creating large single crystals was discovered by the Polish chemist Jan Czochralski in 1916 and is now known as the Czochralski method. To create a single crystal of silicon by using the Czochralski method, electronic-grade silicon (refined to less than one part impurity in 100 billion) is heated to about 1,500 °C (2,700 °F) in a fused quartz cru-cible. Either an electron-donating element such as phosphorus or arsenic (for p-type semiconduc-tors) or an electron-accepting element such as boron (for n-type semiconductors) is mixed in at a concentration of a few parts per billion. A small "seed" crystal, with a diameter of about 0.5 cm (0.2 inch) and a length of about 10 cm (4 inches), is attached to the end of a rod and lowered until it just penetrates the molten surface of the silicon. The rod and the crucible are then rotated in opposite directions while the rod is slowly withdrawn a few millimetres per second. Properly synchronized, these procedures result in the slow growth of a single crystal.

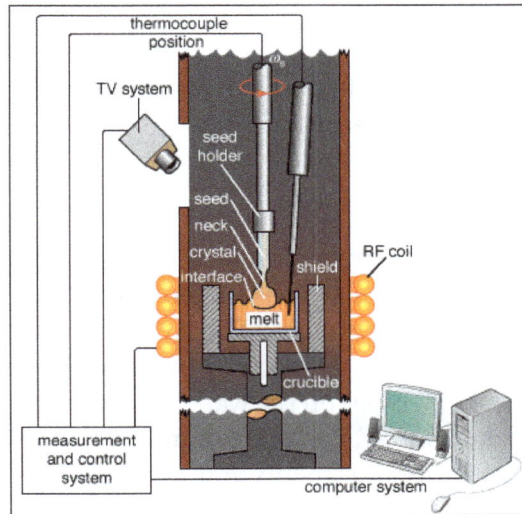

A schematic view of a modern apparatus for crystal pulling using the Czochralski method.

After many days the single crystal can be more than 1 metre (3.3 feet) in length and 300 mm (11.8 inches) in diameter. The large ingot is then sliced like a loaf of bread into thin wafers on which numerous ICs are fabricated simultaneously. The ICs are cut and separated after fabrication.

Using a 0.13-micron process, Intel can produce some 470 Pentium
4 chips from each 300-mm silicon wafer.

Building Layers

All sorts of devices, such as diodes, transistors, capacitors, and resistors, can be built with p- and n-type semiconductors. It is convenient to be able to manufacture all of these different electronic components from the same few basic manufacturing steps.

ICs are made of layers, from about 0.000005 to 0.1 mm thick that are built on the semiconductor substrate one layer at a time, with perhaps 30 or more layers in a final chip. Creating the different electrical components on a chip is a matter of outlining exactly where areas of n- and p-type are to be located on each layer. Each layer is etched, using lines and geometric shapes in the exact locations where the material is to be deposited.

A wafer can be changed in one of three fundamental ways: by deposition (that is, adding a layer), by etching or removing a layer, or by implantation (altering a layer's composition).

Deposition

In a process known as film deposition, a thin film of some substance is deposited onto the wafer by means of either a chemical or a physical reaction.

Chemical Methods

In one common method, known as chemical vapour deposition, the substrate is placed in a low-pressure chamber where certain gases are mixed and heated to 650–850 °C (1,200–1,550 °F) in order to form the desired solid film substance. The solid condenses from the mixed gases and "rains" evenly over the surface of a wafer. A special variant of this technique, known as epitaxy, slowly deposits silicon (or gallium arsenide) on the wafer to produce epitaxial growth of the crystal. Such films can be relatively thick (0.1 mm) and are commonly used for producing silicon-on-insulator substrates that lower the power requirements and speed the switching capabilities of CMOSs. Another variation, known as plasma-enhanced (or plasma-assisted) chemical vapour deposition, uses low pressure and high voltage to create a plasma environment. The plasma causes the gases to react and precipitate at much lower temperatures of 300 to 350 °C (600 to 650 °F) and at faster rates, but this method tends to sacrifice uniformity of deposition.

Two more chemical methods of deposition are electrodeposition (or electroplating) and thermal oxidation. In the former the substrate is given an electrically conducting coating and placed in a liquid solution (electrolyte) containing metal ions, such as gold, copper, or nickel. A wide range of film thicknesses can be built. In thermal oxidation the substrate is heated to 900–1,100 °C (1,650–2,000 °F), which causes the surface to oxidize. This process is often used to form a thin (0.0001-mm) insulating layer of silicon dioxide.

Physical Methods

In general, physical methods of film deposition are less uniform than chemical methods; however, physical methods can be performed at lower temperatures and thus at less risk of damage to the substrate. A common physical method is sputtering. In sputtering, a wafer and a metal source are placed in a vacuum chamber, and an inert gas such as argon is introduced at low pressure. The gas is then ionized by a radio-frequency power source, and the ions are accelerated by an electric field toward the metal surface. When these high-energy ions impact, they knock some of the metal atoms loose from the surface to form a vapour. This vapour condenses on the surfaces within the chamber, including the substrate, where it forms the desired film.

In evaporation deposition, a metal source is heated in a vacuum chamber either by passing a current through a tungsten container or by focusing an electron beam on the metal's surface. As metal atoms evaporate, they form a vapour that condenses on the cooler surface of the wafer to form a layer.

Finally, in casting, a substance is dissolved in a solvent and sprayed on the wafer. After the solvent evaporates, an extremely thin film (perhaps a single layer of molecules) of the substance is left behind. Casting is typically used to add a photosensitive polymer coating, called the photoresist layer.

Etching

A layer can be removed, in entirety or in part, either by etching away the material with strong chemicals or by reactive ion etching (RIE). RIE is like sputtering in the argon chamber, but the polarity is reversed and different gas mixtures are used. The atoms on the surface of the wafer fly away, leaving it bare.

Implantation

Another method of modifying a wafer is to bombard its surface with extra atoms. This is called implantation. Enough of the atoms become deeply embedded in the surface to alter its characteristics, creating areas of n- and p-type materials. Overzealous atoms ripping through the nicely organized crystal lattice damage the structure of the wafer. After implantation the wafer is annealed (heated) to repair this damage. As a side effect of annealing, the implanted atoms usually move a little, diffusing into the surrounding material. The total area that contains implanted atoms after annealing is therefore called a diffusion layer.

A final passivation layer is added to the top of the wafer to seal it from water and other contaminants. Holes are etched through this layer in certain locations to make electrical contact with the integrated circuitry.

Photolithography

In order to alter specific locations on a wafer, a photoresist layer is first applied. Photoresist, or just resist, typically dissolves in a high-pH solution after exposure to light (including ultraviolet radiation or X-rays), and this process, known as development, is controlled by using a mask. A mask is made by applying a thick deposit of chrome in a particular pattern to a glass plate. The chrome provides a shadow over most of the wafer, allowing "light" to shine through only in desired locations. This enables the creation of extremely small areas—depending on the wavelength of the light used—that are unprotected by the hard resist.

After washing away the developed resist, the unprotected areas can be modified through the deposition, etching, or implantation processes, without affecting the rest of the wafer. Once such modifications are finished, the remaining resist is dissolved by a special solvent. This process is repeated with different masks at various layers (30 or so) to create changes to the wafer.

The person who designs the masks for each layer is called the layout engineer, or mask designer. The selection of circuit components and connections is given to mask designers by circuit designers, but mask designers have great latitude in deciding how the end product will be created, which layers will be used to build the components, how to design the connections, how it will look, how large it will be, and how well it will perform. Successful IC development is a team effort between circuit and mask designers.

Final Product

After all the changes to the wafer have been completed, the thousands of individual IC units are sliced apart. This is called dicing the wafer. Each IC unit is now called a die. Dies resemble satellite images of cities, in which circuits look like roadways.

Each die that passes testing is placed into a hard plastic package. These plastic packages, called chips, are what one observes when looking at a computer's circuit board. The plastic packages have metal connection pins that connect the outside world (such as a computer board) to the proper contact points on the die through holes in the passivation layer.

Circuit board showing the microprocessor.

Kirchhoff's Laws

In 1845, a German physicist, Gustav Kirchhoff developed a pair or set of rules or laws which deal with the conservation of current and energy within electrical circuits. These two rules are commonly known as: Kirchhoffs Circuit Laws with one of Kirchhoffs laws dealing with the current flowing around a closed circuit, Kirchhoffs Current Law, (KCL) while the other law deals with the voltage sources present in a closed circuit, Kirchhoffs Voltage Law, (KVL).

Kirchhoffs First Law – The Current Law, (KCL)

Kirchhoffs Current Law or KCL, states that the "total current or charge entering a junction or node is exactly equal to the charge leaving the node as it has no other place to go except to leave, as no charge is lost within the node". In other words the algebraic sum of ALL the currents entering and leaving a node must be equal to zero, I(exiting) + I(entering) = 0. This idea by Kirchhoff is commonly known as the Conservation of Charge.

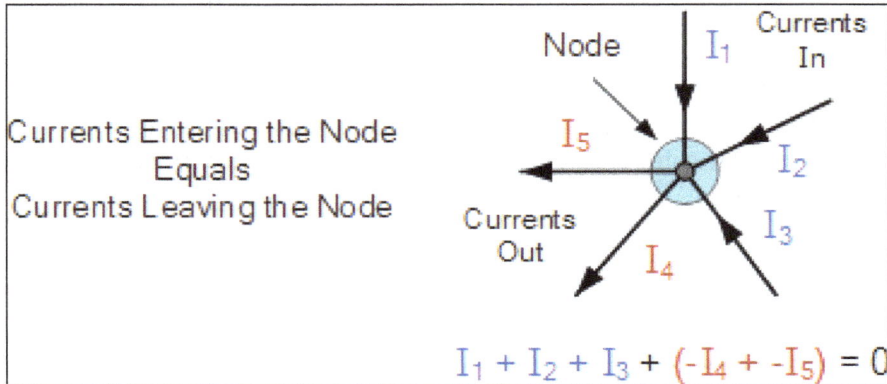

Here, the three currents entering the node, I_1, I_2, I_3 are all positive in value and the two currents leaving the node, I_4 and I_5 are negative in value. Then this means we can also rewrite the equation as;

$$I_1 + I_2 + I_3 - I_4 - I_5 = 0$$

The term Node in an electrical circuit generally refers to a connection or junction of two or more current carrying paths or elements such as cables and components. Also for current to flow either in or out of a node a closed circuit path must exist. We can use Kirchhoff's current law when analysing parallel circuits.

Kirchhoffs Second Law – The Voltage Law, (KVL)

Kirchhoffs Voltage Law or KVL, states that "in any closed loop network, the total voltage around the loop is equal to the sum of all the voltage drops within the same loop" which is also equal to zero. In other words the algebraic sum of all voltages within the loop must be equal to zero. This idea by Kirchhoff is known as the Conservation of Energy.

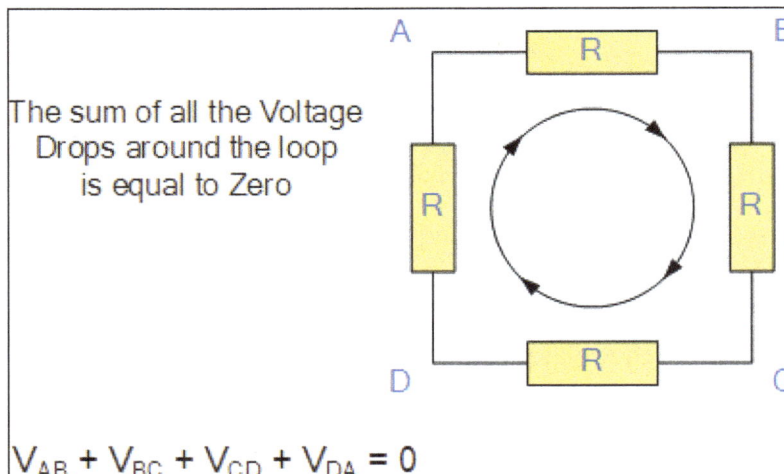

Starting at any point in the loop continue in the same direction noting the direction of all the voltage drops, either positive or negative, and returning back to the same starting point. It is important to maintain the same direction either clockwise or anti-clockwise or the final voltage sum will not be equal to zero. We can use Kirchhoff's voltage law when analysing series circuits.

When analysing either DC circuits or AC circuits using Kirchhoffs Circuit Laws a number of definitions and terminologies are used to describe the parts of the circuit being analysed such as: node, paths, branches, loops and meshes. These terms are used frequently in circuit analysis so it is important to understand them.

Nodal Analysis

Nodal Analysis is used on circuits to obtain multiple KCL equations which are used to solve for voltage and current in a circuit. The number of KCL equations required is one less than the number of nodes that a circuit has. The extra node may be referred to as a reference node. Usually, if a circuit contains a ground, whichever node the ground is connected to is selected as the reference node. This is used to find the voltage differences at each other node in the circuit with respect to the reference.

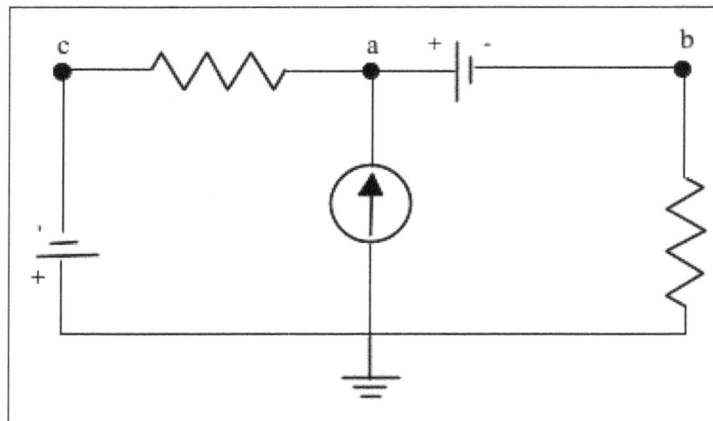

DC circuit showing nodes.

Ideally we set the voltage to 0 V at the reference node to simplify calculations, however it can be set to any value as long as the other nodes account for the different reference voltage. Solving the node equations can provide us with the node voltages.

The node equations are obtained by completing two things:

1.   Express the current through an element in terms of the node voltages.

2.   With the exception of the reference node, apply KCL to each other node in the circuit.

Figure below shows an example of a DC circuit with current and voltage sources. It contains 3 nodes a, b and c, as well as the reference node at the grounded connection.

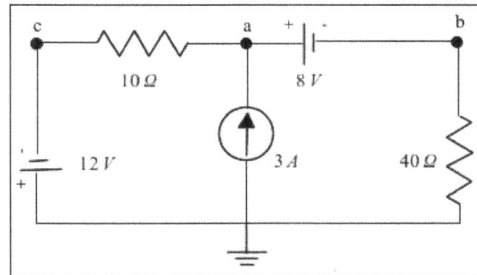

DC Circuit with Voltage and Current Sources.

Here, node c is an example of a supernode which is a connection between two nodes via an in-dependant or dependant voltage source. Because supernodes are connected to a voltage source we can find their voltage immediately. In this case, with the ground at 0 V, the voltage across the source will be $12 = 0 - V_C$, therefore $v_c = -12$ V. Similarily, node a is related to node b as a super-node, $v_a = v_b + 8$ V. We can substitute v_a and v_c into our KCL equations to solve for v_b.

Calculate the KCL equations at node a in figure. The current source is directing current into node a, and we will assume the current flows away from a towards node b and c. The KCL equation for node a is:

$$3\left[\left[A\right]\right] = i_b + i_c$$

The current across the 40Ω resistor represented by the node voltages is found through Ohm's law as the potential across the element divided by the resistance. note the assumed direction of the current to ensure the correct polarity of the difference in potential:

$$i_b = \frac{v_b - 0V}{40\Omega} = \frac{v_b}{40}$$

The current through node c will be equivalent to the potential across the 10Ω resistor divided by the resistance,

$$i_c = \frac{v_a - V_c}{10\Omega}$$

Substituting in the currents i_b and i_c and using the equations for v_a and v_c, the KCL equation at node A is found to be,

$$3 = \frac{v_b}{40} + \frac{\left(v_b + 8\right) - \left(-12\right)}{10}$$

Solving for v_b shows the voltage is 8 volts. Finally, the voltage at a is $v_a = v_b + 8$, so $v_a = 16$ volts.

Mesh Analysis

Mesh Current Analysis

An easier method of solving the above circuit is by using Mesh Current Analysis or Loop Analysis which is also sometimes called Maxwell´s Circulating Currents method. Instead of labelling the branch currents we need to label each "closed loop" with a circulating current.

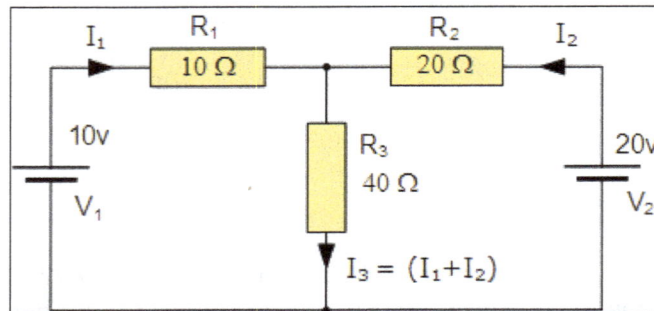

As a general rule of thumb, only label inside loops in a clockwise direction with circulating currents as the aim is to cover all the elements of the circuit at least once. Any required branch current may be found from the appropriate loop or mesh currents as before using Kirchhoff´s method.

For example: $i_1 = I_1$, $i_2 = -I_2$ and $I_3 = I_1 - I_2$

We now write Kirchhoff's voltage law equation in the same way as before to solve them but the advantage of this method is that it ensures that the information obtained from the circuit equations is the minimum required to solve the circuit as the information is more general and can easily be put into a matrix form.

For example, consider the circuit,

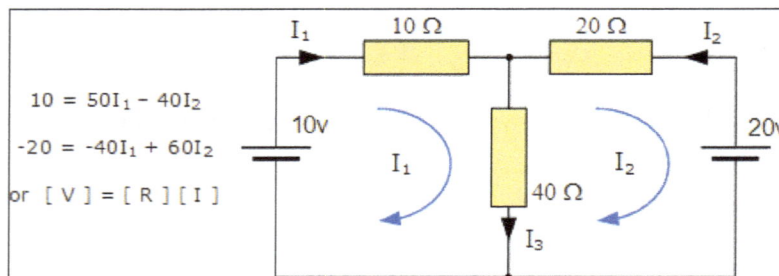

These equations can be solved quite quickly by using a single mesh impedance matrix Z. Each element ON the principal diagonal will be "positive" and is the total impedance of each mesh. Whereas, each element OFF the principal diagonal will either be "zero" or "negative" and represents the circuit element connecting all the appropriate meshes.

First we need to understand that when dealing with matrices, for the division of two matrices it is the same as multiplying one matrix by the inverse of the other as shown:

$$[V] = [I] \times [R] \, or \, [R] \times [I] = [V]$$

$$\begin{bmatrix} 50 & -40 \\ -40 & 60 \end{bmatrix} \times \begin{bmatrix} I_1 \\ I_2 \end{bmatrix} = \begin{bmatrix} 10 \\ -20 \end{bmatrix}$$

$$I = \frac{V}{R} = R^{-1} \times V$$

$$Inverse\ of\ R = \begin{bmatrix} 60 & 40 \\ 40 & 50 \end{bmatrix}$$

$$|R| = (60 \times 50) - (40 \times 40) = 1400$$

$$\therefore R^{-1} = \frac{1}{1400}\begin{bmatrix} 60 & 40 \\ 40 & 50 \end{bmatrix}$$

having found the inverse of R, as V/R is the same as V x R⁻¹, we can now use it to find the two circulating currents.

$$[I] = \left[R^{-1} \right] \times [V]$$

$$\begin{bmatrix} I_1 \\ I_2 \end{bmatrix} = \frac{1}{1400}\begin{bmatrix} 60 & 40 \\ 40 & 50 \end{bmatrix} \times \begin{bmatrix} 10 \\ -20 \end{bmatrix}$$

$$I_1 = \frac{(60 \times 10) + (40 \times -20)}{1400} = \frac{-200}{1400} = -0.143A$$

$$I_2 = \frac{(40 \times 10) + (50 \times -20)}{1400} = -\frac{-600}{1400} = -0.429A$$

Where:

- [V] gives the total battery voltage for loop 1 and then loop 2.

- [I] states the names of the loop currents which we are trying to find.

- [R] is the resistance matrix.

- [R⁻¹] is the inverse of the [R] matrix.

and this gives I_1 as -0.143 Amps and I_2 as -0.429 Amps

As,

$$I_3 = I_1 - I_2$$

The combined current of I_3 is therefore given as: -0.143 − (-0.429) = 0.286 Amps.

Network Theorems

Electric circuit theorems are always beneficial to help find voltage and currents in multi loop circuits. These theorems use fundamental rules or formulas and basic equations of mathematics to analyze basic components of electrical or electronics parameters such as voltages, currents, resistance, and so on. These fundamental theorems include the basic theorems like Superposition theorem, Tellegen's theorem, Norton's theorem, Maximum power transfer theorem and Thevenin's theorems.

Other group of network theorems which are mostly used in the circuit analysis process includes Compensation theorem, Substitution theorem, Reciprocity theorem, Millman's theorem and Miller's theorem.

Super Position Theorem

The Super position theorem is a way to determine the currents and voltages present in a circuit that has multiple sources (considering one source at a time). The super position theorem states that in a linear network having a number of voltage or current sources and resistances, the current through any branch of the network is the algebraic sum of the currents due to each of the sources when acting independently.

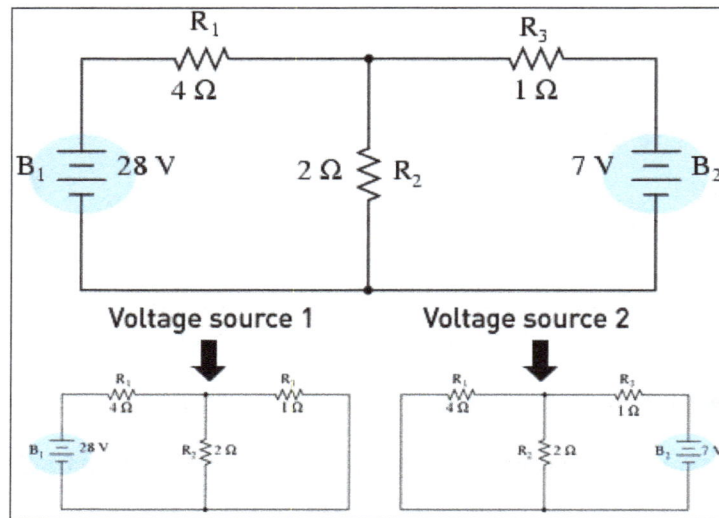

Super Position Theorem.

Super position theorem is used only in linear networks. This theorem is used in both AC and DC circuits wherein it helps to construct Thevenin and Norton equivalent circuit.

In the above figure, the circuit with two voltage sources is divided into two individual circuits according to this theorem's statement. The individual circuits here make the whole circuit look simpler in easier ways. And, by combining these two circuits again after individual simplification, one can easily find parameters like voltage drop at each resistance, node voltages, currents, etc.

Thevenin's Theorem

A linear network consisting of a number of voltage sources and resistances can be replaced by an equivalent network having a single voltage source called Thevenin's voltage (Vthv) and a single resistance called (Rthv).

The below figure explains how this theorem is applicable for circuit analysis. Thevinens voltage is calculated by the given formula between the terminals A and B by breaking the loop at the terminals A and B. Also, Thevinens resistance or equivalent resistance is calculated by shorting voltage sources and open circuiting current sources as shown in the figure.

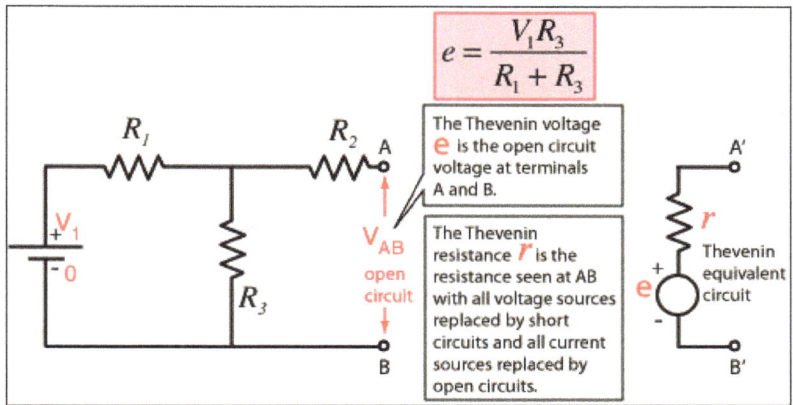

Thevenin's Theorem.

This theorem can be applied to both linear and bilateral networks. It is mainly used for measuring the resistance with a Wheatstone bridge.

Norton's Theorem

This theorem states that any linear circuit containing several energy sources and resistances can be replaced by a single constant current generator in parallel with a single resistor.

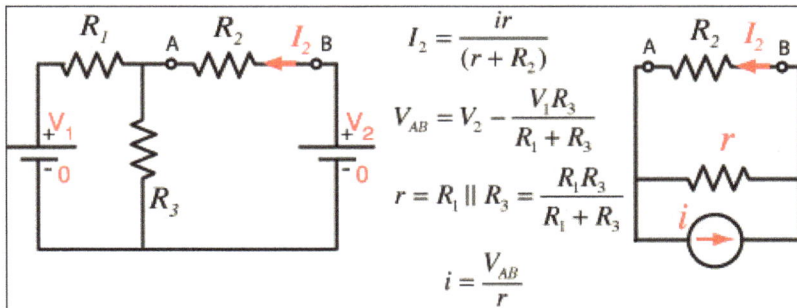

Norton's Theorem.

This is also same as that of the Thevinens theorem, in which we find Thevinens equivalent voltage and resistance values, but here current equivalent values are determined. The process of finding these values is shown as given in the example within the figure.

Maximum Power Transfer Theorem

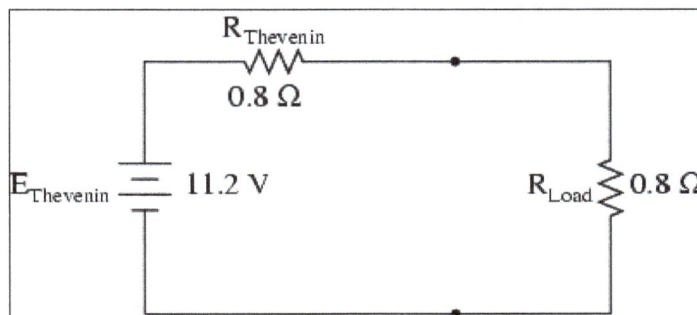

Maximum Power Transfer Theorem.

This theorem explains the condition for the maximum power transfer to load under various circuit conditions. The theorem states that the power transfer by a source to a load is maximum in a network when the load resistance is equal to the internal resistance of the source. For AC circuits load impedance should match with the source impedance for maximum power transfer even if the load is operating at different power factor.

For instance, the above figure depicts a circuit diagram wherein a circuit is simplified up to a level of source with internal resistance using Thevinens theorem. The power transfer will be maximum when this Thevinens resistance is equal to the load resistance. The Practical application of this theorem includes an audio system wherein the resistance of the speaker must be matched to the audio power amplifier to obtain a maximum output.

Reciprocity Theorem

Reciprocity theorem helps to find the other corresponding solution even without further work, once the circuit is analyzed for one solution. The theorem states that in a linear passive bilateral network, the excitation source and its corresponding response can be interchanged.

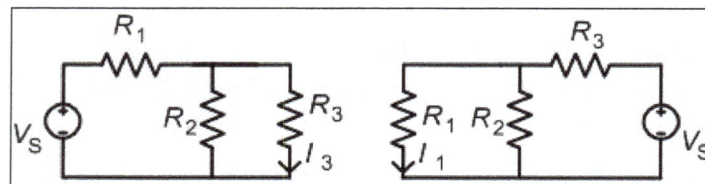

Reciprocity Theorem.

In the above figure, the current in R3 branch is I3 with a single source Vs. If this source is replaced to the R3 branch and shorting the source at the original location, then the current flowing from the original location I1is same as that of I3. This is how we can find corresponding solutions for the circuit once the circuit is analyzed with one solution.

Compensation Theorem

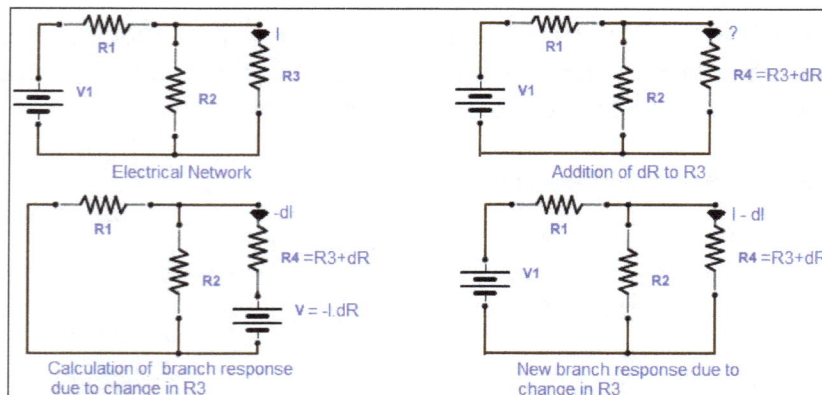

Compensation Theorem.

In any bilateral active network, if the amount of impedance is changed from the original value to some other value carrying a current of I, then the resulting changes that occurs in other branches are same as those that would have been caused by the injection voltage source in the modified branch with a negative sign, i.e., minus of voltage current and changed impedance product. The four figures given above show how this compensation theorem is applicable in analyzing the circuits.

Millman's Theorem

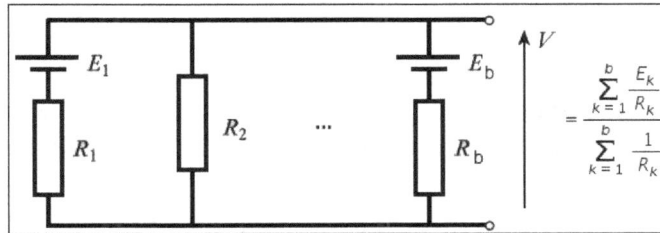

Millman's Theorem.

This theorem states that when any number of voltage sources with finite internal resistance is operating in parallel can be replaced with a single voltage source with series equivalent impedance. The Equivalent voltage for these parallel sources with internal sources in Millmans theorem is calculated by the below given formula, which is shown in the above figure.

Tellegen's Theorem

$$\sum_{k=1}^{n} P_k = V_k \times I_k = 0$$

This theorem is applicable for circuits with a liner or nonlinear, passive or active and hysteric or non-hysteric networks. It states that summation of instantaneous power in circuit with n number of branches is zero.

Substitution Theorem

This theorem states that any branch in a network can be substituted by a different branch without disturbing the currents and voltages in the whole network provided the new branch has the same set of terminal voltages and current as the original branch. Substitution theorem can be used in both linear and nonlinear circuits.

Miller's Theorem

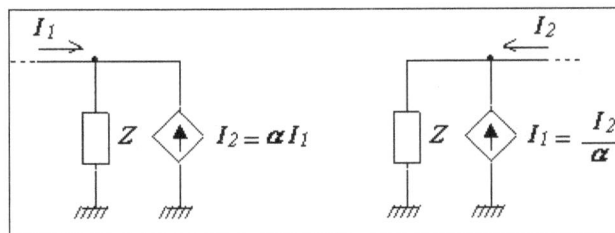

Miller's Theorema.

This theorem states that, in a linear circuit if a branch exists with impedance Z connected between two nodes with nodal voltages, this branch can be replaced by two branches connecting the corresponding nodes to the ground by two impedances. The application of this theorem is not only an effective tool for creating equivalent circuit, but also a tool for designing modified additional electronic circuits by impedance.

DC Circuits

The closed path in which the direct current flows is called the DC circuit. The current flows in only one direction and it is mostly used in low voltage applications. The resistor is the main component of the DC circuit.

A simple DC circuit is shown in the figure which contains a DC source (battery), a load lamp, a switch, connecting leads, and measuring instruments like ammeter and voltmeter. The load resistor is connected in series, parallel or series-parallel combination as per requirement.

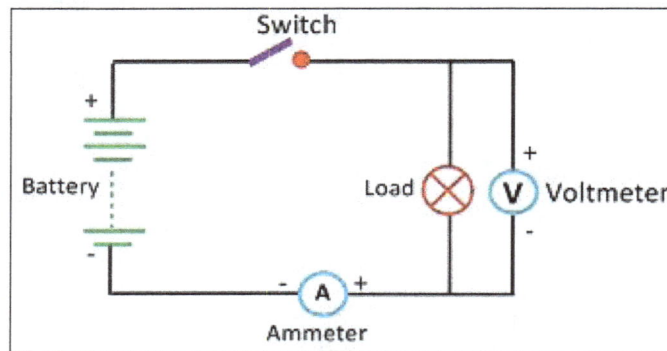

DC Circuit

Types of DC Circuit

The DC electric circuit is mainly classified into three groups. They are the series DC circuit, parallel DC circuit, and series and parallel DC circuit.

DC Series Circuit

The circuit in which have DC series source, and the number of resistors are connected end to end so that same current flow through them is called a DC series circuit. The figure below shows the simple series circuit. In the series circuit the resistor R_1, R_2, and R_3 are connected in series across a supply voltage of V volts. The same current I is flowing through all the three resistors.

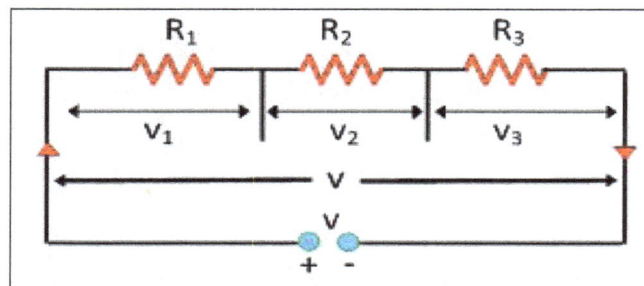

DC Series Circuit

If V_1, V_2, and V_3 are the voltage drop across the three resistor R_1, R_2, and R_3 respectively, then,

$$V = V_1 + V_2 + V_3$$
$$V = IR_1 + IR_2 + IR_3$$

Let R be the total resistance of the circuit then,

$$IR = IR_1 + IR_2 + IR_3$$
$$R = R_1 + R_2 + R_3$$

Total resistance = Sum of the individual resistance.

In such type of circuit all the lamps are controlled by the single switch and they cannot be controlled individually. The most common application of this circuit is for decoration purpose where a number of low voltage lamps are connected in series.

DC Parallel Circuit

The circuit which have DC source and one end of all the resistors is joined to a common point and other end are also joined to another common point so that current flows through them is called a DC parallel circuit.

The figure shows a simple parallel circuit. In this circuit the three resistors R_1, R_2, and R_3 are connected in parallel across a supply voltage of V volts. The current flowing through them is I_1, I_2 and I_3 respectively.

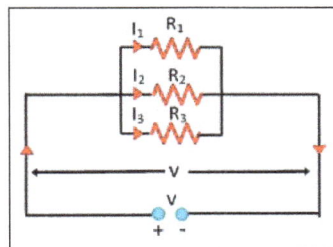

DC Parallel Circuit.

The total current drawn by the circuit,

$$I = I_1 + I_2 + I_3$$
$$I = \frac{V}{R_1} + \frac{V}{R_2} + \frac{V}{R_3}$$

Let R be the total or effective resistance of the circuit, then,

$$\frac{V}{R} = \frac{V}{R_1} + \frac{V}{R_2} + \frac{V}{R_3}$$
$$\frac{1}{R} = \frac{1}{R_1} + \frac{1}{R_2} + \frac{1}{R_3}$$

Reciprocal of total resistance = sum of reciprocal of the individual resistance.

All the resistance is operated to the same voltage; therefore all of them are connected in parallel. Each of them can be controlled individually with the help of a separate switch.

DC Series-Parallel Circuit

The circuit in which series and parallel circuit are connected in series is called a series parallel circuit. The figure shows the series-parallel circuit. In this circuit, two resistors R_1 and R_2 are connected in parallel with each other across terminal AB. The other three resistors R_3, R_4 and R_6 are connected in parallel with each other across terminal BC.

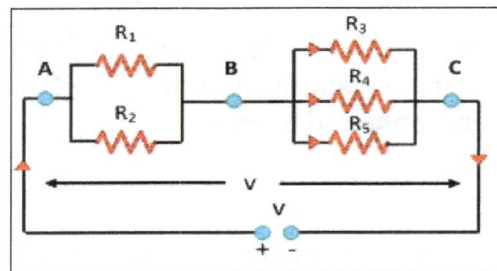

DC Series Parallel Circuit.

The two groups of resistor R_{AB} and R_{BC} are connected in series with each other across the supply voltage of V volts. The total or effective resistance of the whole circuit can be determined as given below,

$$\frac{1}{R_{AB}} = \frac{1}{R_1} + \frac{1}{R_2} = \frac{R_1 + R_2}{R_1 R_2}$$

$$R_{AB} = \frac{R_1 R_2}{R_1 + R_2}$$

Similarly,

$$\frac{1}{R_{BC}} = \frac{1}{R_3} + \frac{1}{R_4} + \frac{1}{R_5} = \frac{R_3 R_4 + R_4 R_5 + R_5 R_3}{R_3 R_4 R_5}$$

$$R_{BC} = \frac{R_3 R_4 R_5}{R_3 R_4 + R_4 R_5 + R_5 R_3}$$

Total or effective resistance of the ciruit,

$$R = R_{AB} + R_{BC}$$

AC Circuits

The path for the flow of alternating current is called an AC Circuit. The alternating current (AC) is used for domestic and industrial purposes. In an AC circuit, the value of the magnitude and the direction of current and voltages are not constant; it changes at a regular interval of time. It travels as a sinusoidal wave completing one cycle as half positive and half negative cycle and is a function of time (t) or angle (θ = wt).

In DC Circuit, the opposition to the flow of current is the only resistance of the circuit whereas the opposition to the flow of current in the AC circuit is because of resistance (R), Inductive Reactance ($X_L = 2\pi fL$) and capacitive reactance ($X_C = 1/2\ \pi fC$) of the circuit.

In AC Circuit, the current and voltages are represented by magnitude and direction. The alternating quantity may or may not be in phase with each other depending upon the various parameters of the circuit like resistance, inductance, and capacitance. The sinusoidal alternating quantities are voltage and current which varies according to the sine of angle θ.

For the generation of electric power, in all over the world the sinusoidal voltage and current are selected because of the following reasons are given below.

- The sinusoidal voltage and current produce low iron and copper losses in the transformer and rotating electrical machines, which in turns improves the efficiency of the AC machines.

- They offer less interference to the nearby communication system.

- They produce fewer disturbances in the electrical circuit.

Alternating Voltage and Current in an AC Circuit

The voltage that changes its polarity and magnitude at regular interval of time is called an alternating voltage. Similarly the direction of the current is changed and the magnitude of current changes with time it is called alternating current. When an alternating voltage source is connected across a load resistance as shown in the figure below, the current through it flows in one direction and then in the opposite direction when the polarity is reversed.

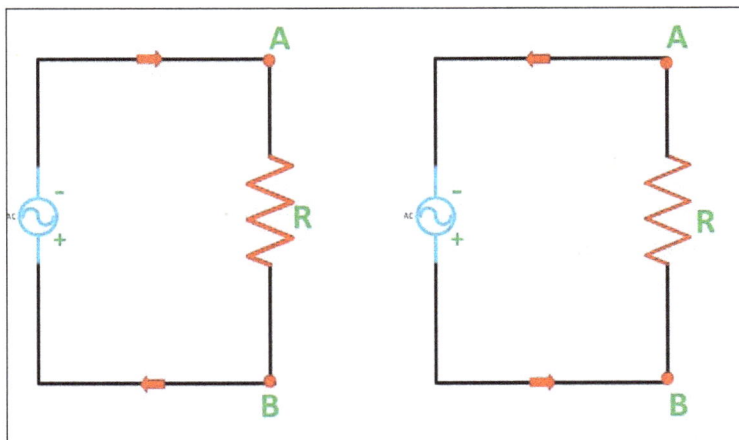

Alternating Currnt Circuit Diagram.

The waveform of the alternating voltage with respect to the time and the current flowing through the resistance (R) in the circuit is shown in figure.

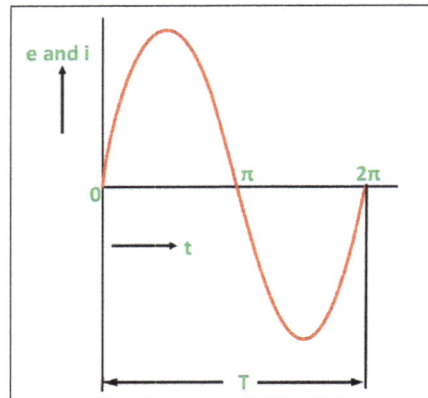

There are various types of AC circuit such as AC circuit containing only resistance (R), AC circuit containing only capacitance (C), AC circuit containing only inductance (L), the combination of RL Circuit, AC circuit containing resistance and capacitance (RC), AC circuit containing inductance and capacitance (LC) and resistance inductance and capacitance (RLC) AC circuit.

The various terms which are frequently used in an AC Circuit are as follows:

- Amplitude: The maximum positive or negative value attained by an alternating quantity in one complete cycle is called Amplitude or peak value or maximum value. The maximum value of voltage and current is represented by E_m or V_m and I_m respectively.

- Alternation: One half cycles is termed as alternation. An alternation span is of 180 degrees electrical.

- Cycle: When one set of positive and negative values completes by an alternating quantity or it goes through 360 degrees electrical, it is said to have one complete Cycle.

- Instantaneous Value: The value of voltage or current at any instant of time is called an instantaneous value. It is denoted by (i or e).

- Frequency: The number of cycles made per second by an alternating quantity is called frequency. It is measured in cycle per second (c/s) or hertz (Hz) and is denoted by (f).

- Time Period: The time taken in seconds by a voltage or a current to complete one cycle is called Time Period. It is denoted by (T).

- Wave Form: The shape obtained by plotting the instantaneous values of an alternating quantity such as voltage and current along the y axis and the time (t) or angle ($\theta = wt$) along the x axis is called waveform.

Resistance in an AC circuit

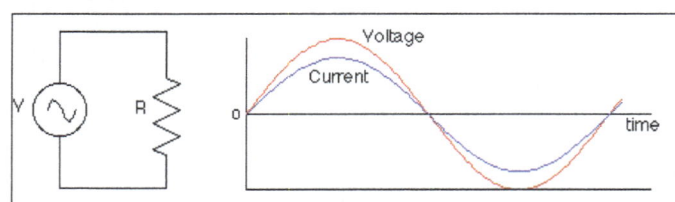

The relationship V = IR applies for resistors in an AC circuit, so,

$$I = V / R = \left(V_0 / R\right)\sin\left(\omega t\right) = I_0 \sin\left(\omega t\right)$$

In AC circuits we'll talk a lot about the phase of the current relative to the voltage. In a circuit which only involves resistors, the current and voltage are in phase with each other, which means that the peak voltage is reached at the same instant as peak current. In circuits which have capacitors and inductors (coils) the phase relationships will be quite different.

Capacitance in an AC Circuit

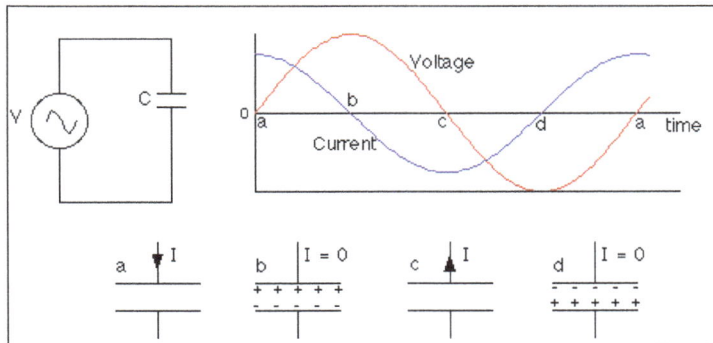

Consider now a circuit which has only a capacitor and an AC power source (such as a wall outlet). A capacitor is a device for storing charging. It turns out that there is a 90° phase difference between the current and voltage, with the current reaching its peak 90° (1/4 cycle) before the voltage reaches its peak. Put another way, the current leads the voltage by 90° in a purely capacitive circuit.

To understand why this is, we should review some of the relevant equations, including:

- Relationship between voltage and charge for a capacitor: CV = Q.

- Relationship between current and the flow charge: I = $\Delta Q / \Delta t$.

The AC power supply produces an oscillating voltage. We should follow the circuit through one cycle of the voltage to figure out what happens to the current.

- Step 1 - At point a the voltage is zero and the capacitor is uncharged. Initially, the voltage increases quickly. The voltage across the capacitor matches the power supply voltage, so the current is large to build up charge on the capacitor plates. The closer the voltage gets to its peak, the slower it changes, meaning less current has to flow. When the voltage reaches a peak at point b, the capacitor is fully charged and the current is momentarily zero.

- Step 2 - After reaching a peak, the voltage starts dropping. The capacitor must discharge now, so the current reverses direction. When the voltage passes through zero at point c, it's changing quite rapidly; to match this voltage the current must be large and negative.

- Step 3 - Between points c and d, the voltage is negative. Charge builds up again on the capacitor plates, but the polarity is opposite to what it was in step one. Again the current is negative, and as the voltage reaches its negative peak at point d the current drops to zero.

- Step 4 - After point d, the voltage heads toward zero and the capacitor must discharge. When the voltage reaches zero it's gone through a full cycle so it's back to point a again to repeat the cycle.

The larger the capacitance of the capacitor, the more charge has to flow to build up a particular voltage on the plates, and the higher the current will be. The higher the frequency of the voltage, the shorter the time available to change the voltage, so the larger the current has to be. The current, then, increases as the capacitance increases and as the frequency increases.

Usually this is thought of in terms of the effective resistance of the capacitor, which is known as the capacitive reactance, measured in ohms. There is an inverse relationship between current and resistance, so the capacitive reactance is inversely proportional to the capacitance and the frequency:

A capacitor in an AC circuit exhibits a kind of resistance called capacitive reactance, measured in ohms. This depends on the frequency of the AC voltage, and is given by,

Capacitive reactance : $X_c = 1 / \omega C = 1 / 2 \pi fc$

We can use this like a resistance (because, really, it is a resistance) in an equation of the form $V = IR$ to get the voltage across the capacitor:

$V = I X_c$

Note that V and I are generally the rms values of the voltage and current.

Inductance in an AC Circuit

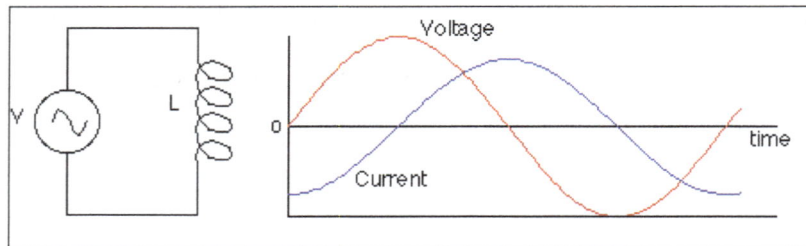

An inductor is simply a coil of wire (often wrapped around a piece of ferromagnet). If we now look at a circuit composed only of an inductor and an AC power source, we will again find that there is a 90° phase difference between the voltage and the current in the inductor. This time, however, the current lags the voltage by 90°, so it reaches its peak 1/4 cycle after the voltage peaks.

The reason for this has to do with the law of induction:

$\varepsilon = -N\Delta\Phi / \Delta t$ or $\varepsilon = -L\Delta I / \Delta t$

Applying Kirchoff's loop rule to the circuit above gives:

$Y - L\Delta I / \Delta t = 0$ so $Y = L\Delta I / \Delta t$

As the voltage from the power source increases from zero, the voltage on the inductor matches it. With the capacitor, the voltage came from the charge stored on the capacitor plates (or, equivalently, from the electric field between the plates). With the inductor, the voltage comes from changing the flux through the coil, or, equivalently, changing the current through the coil, which changes the magnetic field in the coil.

To produce a large positive voltage, a large increase in current is required. When the voltage passes through zero, the current should stop changing just for an instant. When the voltage is large and negative, the current should be decreasing quickly. These conditions can all be satisfied by having the current vary like a negative cosine wave, when the voltage follows a sine wave.

How does the current through the inductor depend on the frequency and the inductance? If the frequency is raised, there is less time to change the voltage. If the time interval is reduced, the change in current is also reduced, so the current is lower. The current is also reduced if the inductance is increased.

As with the capacitor, this is usually put in terms of the effective resistance of the inductor. This effective resistance is known as the inductive reactance. This is given by:

$$X_L = \omega L = 2\pi f\, L$$

where L is the inductance of the coil (this depends on the geometry of the coil and whether it's got a ferromagnetic core). The unit of inductance is the henry.

As with capacitive reactance, the voltage across the inductor is given by:

$$V = IX_L$$

Electrical Energy in an AC Circuit

One of the main differences between resistors, capacitors, and inductors in AC circuits is in what happens with the electrical energy. With resistors, power is simply dissipated as heat. In a capacitor, no energy is lost because the capacitor alternately stores charge and then gives it back again. In this case, energy is stored in the electric field between the capacitor plates. The amount of energy stored in a capacitor is given by:

Energy in a capacitor: Energy = 1/2 CV^2

In other words, there is energy associated with an electric field. In general, the energy density (energy per unit volume) in an electric field with no dielectric is:

Energy density in an electric field = 1/2 $\varepsilon_0 E^2$

With a dielectric, the energy density is multiplied by the dielectric constant.

There is also no energy lost in an inductor, because energy is alternately stored in the magnetic field and then given back to the circuit. The energy stored in an inductor is:

Energy in an inductor: Energy = 1/2 LI^2

Again, there is energy associated with the magnetic field. The energy density in a magnetic field is:

Energy density in a magnetic field = $B^2 / (2\mu_0)$

RLC Circuits

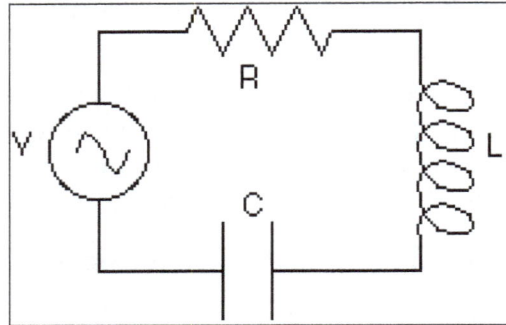

Consider what happens when resistors, capacitors, and inductors are combined in one circuit. If all three components are present, the circuit is known as an RLC circuit (or LRC). If only two components are present, it's either an RC circuit, an RL circuit, or an LC circuit.

The overall resistance to the flow of current in an RLC circuit is known as the impedance, symbolized by Z. The impedance is found by combining the resistance, the capacitive reactance, and the inductive reactance. Unlike a simple series circuit with resistors, however, where the resistances are directly added, in an RLC circuit the resistance and reactances are added as vectors.

This is because of the phase relationships. In a circuit with just a resistor, voltage and current are in phase. With only a capacitor, current is 90° ahead of the voltage and with just an inductor the reverse is true, the voltage leads the current by 90°. When all three components are combined into one circuit, there has to be some compromise.

To figure out the overall effective resistance, as well as to determine the phase between the voltage and current, the impedance is calculated like this. The resistance R is drawn along the +x-axis of an x-y coordinate system. The inductive reactance is at 90° to this, and is drawn along the +y-axis. The capacitive reactance is also at 90° to the resistance, and is 180° different from the inductive reactance, so it's drawn along the -y-axis. The impedance, Z, is the sum of these vectors, and is given by:

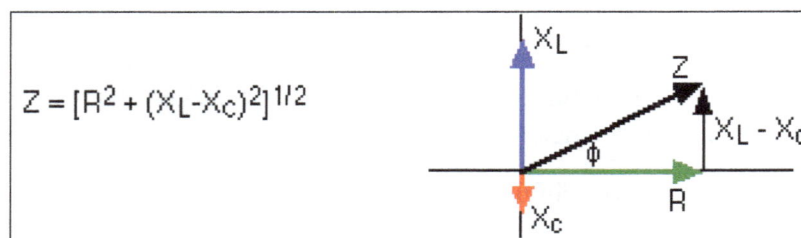

$$Z = [R^2 + (X_L - X_C)^2]^{1/2}$$

The current and voltage in an RLC circuit are related by V = IZ. The phase relationship between the current and voltage can be found from the vector diagram: It's the angle between the impedance, Z, and the resistance, R. The angle can be found from:

$$\tan \Phi = (X_L - X_C)/ R$$

If the angle is positive, the voltage leads the current by that angle. If the angle is negative, the voltage lags the currents.

The power dissipated in an RLC circuit is given by:

P = VI cos Φ (cos Φ is known as the power factor in the circuit)

Note that all of this power is lost in the resistor; the capacitor and inductor alternately store energy in electric and magnetic fields and then give that energy back to the circuit.

Power Factor

In electrical engineering, power factor is only and only related to AC circuits i.e. there is no power factor (P. f) in DC circuits due to zero frequency and phase angle difference (Φ) between current and voltage.

In AC circuits, the power factor is the ratio of the real power that is used to do work and the apparent power that is supplied to the circuit.

The power factor can get values in the range from 0 to 1.

When all the power is reactive power with no real power (usually inductive load) - the power factor is 0.

When all the power is real power with no reactive power (resistive load) - the power factor is 1.

The power factor is equal to the real or true power P in watts (W) divided by the apparent power |S| in volt-ampere (VA):

$$PF = P_{(W)} \, / \, | S_{(VA)} |$$

Where,

PF - Power factor.

P - Real power in watts (W).

|S|- apparent power - the magnitude of the complex power in volt·amps (VA).

Power Factor Calculations

For sinusuidal current, the power factor PF is equal to the absolute value of the cosine of the apparent power phase angle φ (which is also is impedance phase angle):

$$PF = \, | \cos \varphi |$$

PF is the power factor.

φ is the apprent power phase angle.

The real power P in watts (W) is equal to the apparent power |S| in volt-ampere (VA) times the power factor PF:

$$P_{(W)} = | S_{(VA)} | \, \times PF = | S_{(VA)} | \times | \cos \varphi |$$

When the circuit has a resistive impedance load, the real power P is equal to the apparent power |S| and the power factor PF is equal to 1:

$$PF_{(\text{resistive load})} = P / |S| = 1$$

The reactive power Q in volt-amps reactive (VAR) is equal to the apparent power |S| in volt-ampere (VA) times the sine of the phase angle φ:

$$Q_{(\text{VAR})} = |S_{(\text{VA})}| \times |\sin\varphi|$$

Single phase circuit calculation from real power meter reading P in kilowatts (kW), voltage V in volts (V) and current I in amps (A):

$$PF = |\cos\varphi| = 1000 \times P_{(\text{kW})} / (V_{(\text{V})} \times I_{(\text{A})})$$

Three phase circuit calculation from real power meter reading P in kilowatts (kW), line to line voltage $V_{\text{L-L}}$ in volts (V) and current I in amps (A):

$$PF = |\cos\varphi| = 1000 \times P_{(\text{kW})} / (\sqrt{3} \times V_{\text{L-L(V)}} \times I_{(\text{A})})$$

Three phase circuit calculation from real power meter reading P in kilowatts (kW), line to line neutral $V_{\text{L-N}}$ in volts (V) and current I in amps (A):

$$PF = |\cos\varphi| = 1000 \times P_{(\text{kW})} / (3 \times V_{\text{L-N(V)}} \times I_{(\text{A})}).$$

Power Factor Correction

Power factor correction is an adjustment of the electrical circuit in order to change the power factor near 1.

Power factor near 1 will reduce the reactive power in the circuit and most of the power in the circuit will be real power. This will also reduce power lines losses.

The power factor correction is usually done by adding capacitors to the load circuit, when the circuit has inductive components, like an electric motor.

Power Factor Correction Calculation

The apparent power |S| in volt-amps (VA) is equal to the voltage V in volts (V) times the current I in amps (A):

$$|S_{(\text{VA})}| = V_{(\text{V})} \times I_{(\text{A})}$$

The reactive power Q in volt-amps reactive (VAR) is equal to the square root of the square of the apparent power |S| in volt-ampere (VA) minus the square of the real power P in watts (W) (Pythagorean Theorem):

$$Q_{(\text{VAR})} = \sqrt{\left(|S_{(\text{VA})}|^2 - P_{(\text{W})}^2\right)}$$

$$Q_{c\,(\text{kVAR})} = Q_{(\text{kVAR})} - Q_{\text{corrected}\,(\text{kVAR})}$$

The reactive power Q in volt-amps reactive (VAR) is equal to the square of voltage V in volts (V) divided by the reactance Xc:

$$Q_{c\,(VAR)} = V_{(V)}^2 / X_c = V_{(V)}^2 / (1/(2\pi f_{(Hz)} \times C_{(F)}))$$
$$= 2\pi f_{(Hz)} \times C_{(F)} \times V_{(V)}^2$$

So, the power factor correction capacitor in Farad (F) that should be added to the circuit in parallel is equal to the reactive power Q in volt-amps reactive (VAR) divided by 2π times the frequency f in Hertz (Hz) times the squared voltage V in volts (V):

$$C_{(F)} = Q_{c\,(VAR)} / (2\pi f_{(Hz)} \cdot V_{(V)}^2)$$

Linear Circuits

In a purely resistive AC circuit, voltage and current waveforms are in step (or in phase), changing polarity at the same instant in each cycle. All the power entering the load is consumed (or dissipated).

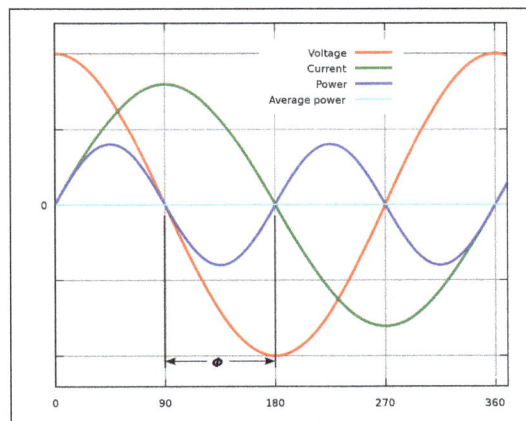

Instantaneous and average power calculated from AC voltage and current with a zero power factor (φ = 90°, cos(φ) = 0). The blue line shows all the power is stored temporarily in the load during the first quarter cycle and returned to the grid during the second quarter cycle, so no real power is consumed.

Where reactive loads are present, such as with capacitors or inductors, energy storage in the loads results in a phase difference between the current and voltage waveforms. During each cycle of the AC voltage, extra energy, in addition to any energy consumed in the load, is temporarily stored in the load in electric or magnetic fields, and then returned to the power grid a fraction of the period later.

In the electric power grid, reactive loads thus cause a continuous "ebb and flow" of nonproductive power. A circuit with a low power factor will use higher currents to transfer a given quantity of real power than a circuit with a high power factor, causing increased losses due to resistive heating in power lines, and requiring higher rated conductors and transformers be installed. A linear load does not change the shape of the waveform of the current, but may change the relative timing (phase) between voltage and current.

Electrical circuits containing dominantly resistive loads (incandescent lamps, heating elements) have a power factor of almost 1.0, but circuits containing inductive or capacitive loads (electric motors, solenoid valves, transformers, fluorescent lamp ballasts, and others) can have a power factor well below 1.

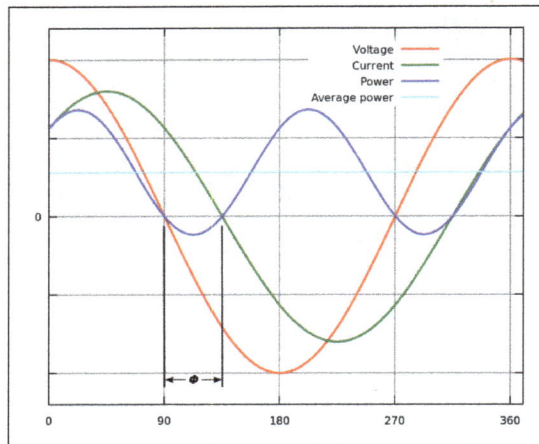

Instantaneous and average power calculated from AC voltage and current with a lagging power factor ($\varphi = 45°$, $\cos(\varphi) \approx 0.71$). The blue line shows some of the power is returned to the grid during the part of the cycle labeled φ.

Definition and Calculation

AC power flow has two components:

- Real power or active power (P) (sometimes called average power), expressed in watts (W).

- Reactive power (Q), usually expressed in reactive volt-amperes (var).

These are combined to the complex power (S) expressed volt-amperes (VA). The magnitude of the complex power is the apparent power (S), also expressed in volt-amperes (VA).

The VA and var are non-SI units mathematically identical to the watt, but are used in engineering practice instead of the watt to state what quantity is being expressed. The SI explicitly disallows using units for this purpose or as the only source of information about a physical quantity as used.

The power factor is defined as the ratio of real power to apparent power. As power is transferred along a transmission line, it does not consist purely of real power that can do work once transferred to the load, but rather consists of a combination of real and reactive power, called apparent power. The power factor describes the amount of real power transmitted along a transmission line relative to the total apparent power flowing in the line.

Power Triangle

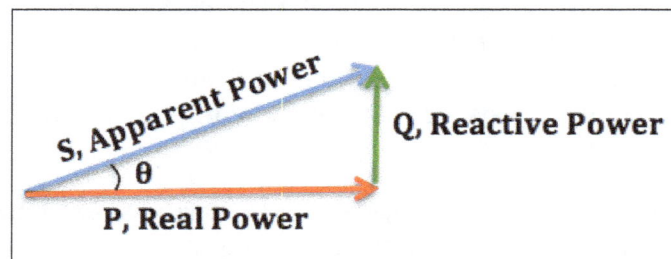

One can relate the various components of AC power by using the power triangle in vector space. Real power extends horizontally in the î direction as it represents a purely real component of AC power. Reactive power extends in the direction of ĵ as it represents a purely imaginary component of AC power. Complex power (and its magnitude, Apparent power) represents a combination of both real and reactive power, and therefore can be calculated by using the vector sum of these two components. We can conclude that the mathematical relationship between these components is:

$$S = P + jQ$$
$$|S|^2 = P^2 + Q^2$$
$$|S| = \sqrt{P^2 + Q^2}$$
$$\cos\theta, \text{ power factor} = \frac{P, \text{ real power}}{|S|, \text{ apparent power}}$$

or

$$\text{power factor} = \cos(\arctan(Q/P))$$

it follows that:

$$Q = P * \tan(\arccos(\text{power factor}))$$

Increasing the Power Factor

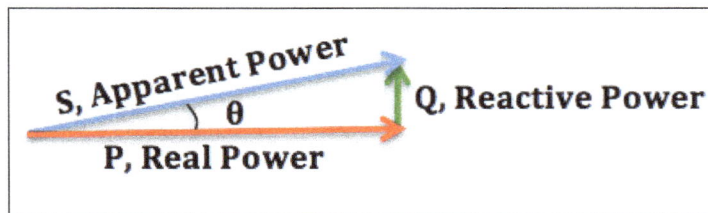

As the power factor (i.e. $\cos\theta$) increases, the ratio of real power to apparent power (which = $\cos\theta$), increases and approaches unity (1), while the angle θ decreases and the reactive power decreases. [As $\cos\theta \to 1$, its maximum possible value, $\theta \to 0$ and so $Q \to 0$, as the load becomes less reactive and more purely resistive].

Decreasing the Power Factor

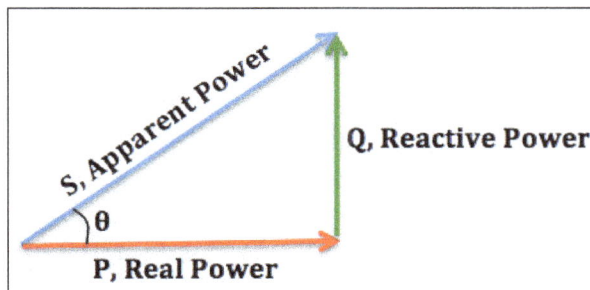

As the power factor decreases, the ratio of real power to apparent power also decreases, as the angle θ increases and reactive power increases.

Lagging and Leading Power Factors

There is also a difference between a lagging and leading power factor. The terms refer to whether the phase of the current is leading or lagging the phase of the voltage. A lagging power factor signifies that the load is inductive, as the load will "consume" reactive power, and therefore the reactive component Q is positive as reactive power travels through the circuit and is "consumed" by the inductive load. A leading power factor signifies that the load is capacitive, as the load "supplies" reactive power, and therefore the reactive component Q is negative as reactive power is being supplied to the circuit.

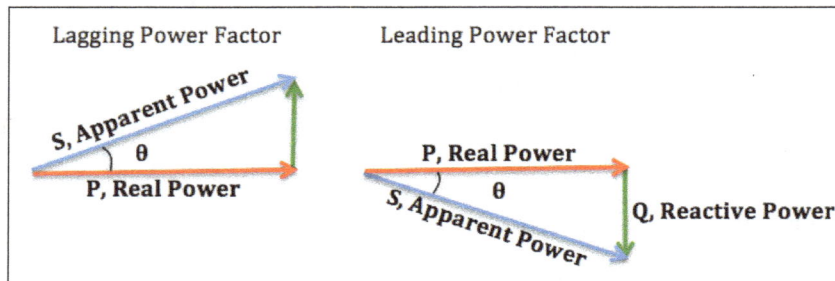

If θ is the phase angle between the current and voltage, then the power factor is equal to the cosine of the angle, $\cos\theta$ ::

$$|P| = |S| \cos\theta$$

Since the units are consistent, the power factor is by definition a dimensionless number between −1 and 1. When power factor is equal to 0, the energy flow is entirely reactive and stored energy in the load returns to the source on each cycle. When the power factor is 1, all the energy supplied by the source is consumed by the load. Power factors are usually stated as "leading" or "lagging" to show the sign of the phase angle. Capacitive loads are leading (current leads voltage), and inductive loads are lagging (current lags voltage).

If a purely resistive load is connected to a power supply, current and voltage will change polarity in step, the power factor will be 1, and the electrical energy flows in a single direction across the network in each cycle. Inductive loads such as induction motors (any type of wound coil) consume reactive power with current waveform lagging the voltage. Capacitive loads such as capacitor banks or buried cable generate reactive power with current phase leading the voltage. Both types of loads will absorb energy during part of the AC cycle, which is stored in the device's magnetic or electric field, only to return this energy back to the source during the rest of the cycle.

For example, to get 1 kW of real power, if the power factor is unity, 1 kVA of apparent power needs to be transferred (1 kW ÷ 1 = 1 kVA). At low values of power factor, more apparent power needs to be transferred to get the same real power. To get 1 kW of real power at 0.2 power factor, 5 kVA of apparent power needs to be transferred (1 kW ÷ 0.2 = 5 kVA). This apparent power must be produced and transmitted to the load, and is subject to the losses in the production and transmission processes.

Electrical loads consuming alternating current power consume both real power and reactive power. The vector sum of real and reactive power is the apparent power. The presence of reactive power causes the real power to be less than the apparent power, and so, the electric load has a power factor of less than 1.

A negative power factor (0 to −1) can result from returning power to the source, such as in the case of a building fitted with solar panels when surplus power is fed back into the supply.

Power Factor Correction of Linear Loads

A high power factor is generally desirable in a power delivery system to reduce losses and improve voltage regulation at the load. Compensating elements near an electrical load will reduce the apparent power demand on the supply system. Power factor correction may be applied by an electric power transmission utility to improve the stability and efficiency of the network. Individual electrical customers who are charged by their utility for low power factor may install correction equipment to increase their power factor so as to reduce costs.

Power factor correction brings the power factor of an AC power circuit closer to 1 by supplying or absorbing reactive power, adding capacitors or inductors that act to cancel the inductive or capacitive effects of the load, respectively. In the case of offsetting the inductive effect of motor loads, capacitors can be locally connected. These capacitors help to generate reactive power to meet the demand of the inductive loads. This will keep that reactive power from having to flow all the way from the utility generator to the load. In the electricity industry, inductors are said to consume reactive power and capacitors are said to supply it, even though reactive power is just energy moving back and forth on each AC cycle.

The reactive elements in power factor correction devices can create voltage fluctuations and harmonic noise when switched on or off. They will supply or sink reactive power regardless of whether there is a corresponding load operating nearby, increasing the system's no-load losses. In the worst case, reactive elements can interact with the system and with each other to create resonant conditions, resulting in system instability and severe overvoltage fluctuations. As such, reactive elements cannot simply be applied without engineering analysis.

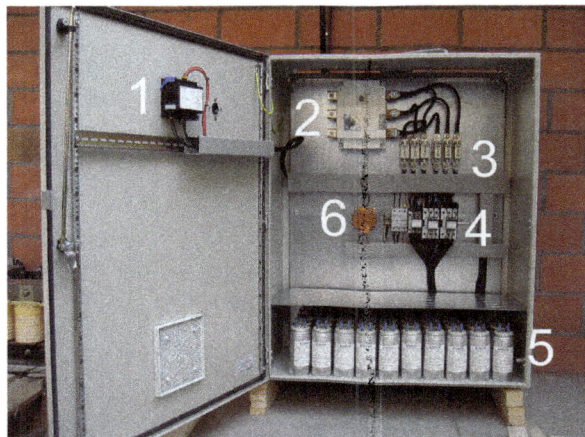

1. Reactive Power Control Relay; 2. Network connection points; 3. Slow-blow Fuses; 4. Inrush Limiting Contactors; 5. Capacitors (single-phase or three-phase units, delta-connection); 6. Transformer (for controls and ventilation fans).

An automatic power factor correction unit consists of a number of capacitors that are switched by means of contactors. These contactors are controlled by a regulator that measures power factor in an electrical network. Depending on the load and power factor of the network, the power factor controller will switch the necessary blocks of capacitors in steps to make sure the power factor stays above a selected value.

In place of a set of switched capacitors, an unloaded synchronous motor can supply reactive power. The reactive power drawn by the synchronous motor is a function of its field excitation. It is referred to as a synchronous condenser. It is started and connected to the electrical network. It operates at a leading power factor and puts vars onto the network as required to support a system's voltage or to maintain the system power factor at a specified level.

The synchronous condenser's installation and operation are identical to those of large electric motors. Its principal advantage is the ease with which the amount of correction can be adjusted; it behaves like a variable capacitor. Unlike with capacitors, the amount of reactive power supplied is proportional to voltage, not the square of voltage; this improves voltage stability on large networks. Synchronous condensers are often used in connection with high-voltage direct-current transmission projects or in large industrial plants such as steel mills.

For power factor correction of high-voltage power systems or large, fluctuating industrial loads, power electronic devices such as the Static VAR compensator or STATCOM are increasingly used. These systems are able to compensate sudden changes of power factor much more rapidly than contactor-switched capacitor banks, and being solid-state require less maintenance than synchronous condensers.

Non-linear Loads

Examples of non-linear loads on a power system are rectifiers (such as used in a power supply), and arc discharge devices such as fluorescent lamps, electric welding machines, or arc furnaces. Because current in these systems is interrupted by a switching action, the current contains frequency components that are multiples of the power system frequency. Distortion power factor is a measure of how much the harmonic distortion of a load current decreases the average power transferred to the load.

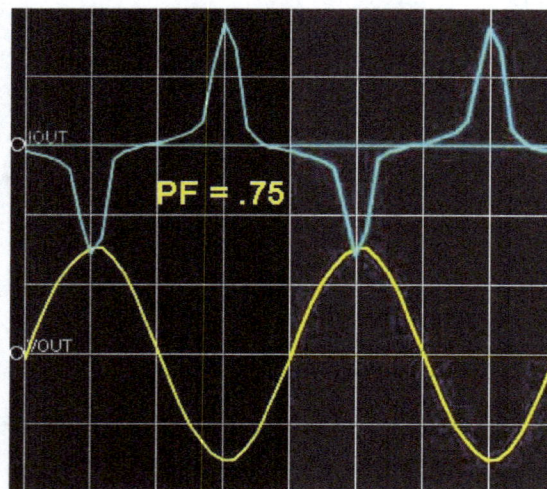

Sinusoidal voltage and non-sinusoidal current give a distortion power factor of 0.75 for this computer power supply load.

Non-sinusoidal Components

In linear circuits having only sinusoidal currents and voltages of one frequency, the power factor arises only from the difference in phase between the current and voltage. This is "displacement power factor".

Non-linear loads change the shape of the current waveform from a sine wave to some other form. Non-linear loads create harmonic currents in addition to the original (fundamental frequency) AC current. This is of importance in practical power systems that contain non-linear loads such as rectifiers, some forms of electric lighting, electric arc furnaces, welding equipment, switched-mode power supplies, variable speed drives and other devices. Filters consisting of linear capacitors and inductors can prevent harmonic currents from entering the supplying system.

To measure the real power or reactive power, a wattmeter designed to work properly with non-sinusoidal currents must be used.

Distortion Power Factor

The *distortion power factor* is the distortion component associated with the harmonic voltages and currents present in the system:

$$\text{distortion power factor} = \frac{I_1}{I_{rms}}$$

$$= \frac{I_1}{\sqrt{I_1^2 + I_2^2 + I_3^2 + I_4^2 + \cdots}}$$

$$= \frac{1}{\sqrt{1 + \dfrac{I_2^2 + I_3^2 + I_4^2 + \cdots}{I_1^2}}}$$

$$= \frac{1}{\sqrt{1 + THD_i^2}}$$

THD_i is the total harmonic distortion of the load current,

$$THD_i = \frac{\sqrt{\sum_{h=2}^{\infty} I_h^2}}{I_1} = \frac{\sqrt{I_2^2 + I_3^2 + I_4^2 + \cdots}}{I_1}$$

I_1 is the fundamental component of the current and I_{rms} is the total current – both are root mean square-values (distortion power factor can also be used to describe individual order harmonics, using the corresponding current in place of total current). This definition with respect to total harmonic distortion assumes that the voltage stays undistorted (sinusoidal, without harmonics). This simplification is often a good approximation for stiff voltage sources (not being affected by changes in load downstream in the distribution network). Total harmonic distortion of typical generators from current distortion in the network is on the order of 1–2%, which can have larger scale implications but can be ignored in common practice.

The result when multiplied with the displacement power factor (DPF) is the overall, true power factor or just power factor (PF):

$$PF = \frac{\cos\varphi}{\sqrt{1 + THD_i^2}}$$

Distortion in Three-phase Networks

In practice, the local effects of distortion current on devices in a three-phase distribution network rely on the magnitude of certain order harmonics rather than the total harmonic distortion.

For example, the triplen, or zero-sequence, harmonics (3rd, 9th, 15th, etc.) have the property of being in-phase when compared line-to-line. In a delta-wye transformer, these harmonics can result in circulating currents in the delta windings and result in greater resistive heating. In a wye-configuration of a transformer, triplen harmonics will not create these currents, but they will result in a non-zero current in the neutral wire. This could overload the neutral wire in some cases and create error in kilowatt-hour metering systems and billing revenue. The presences of current harmonics in a transformer also result in larger eddy currents in the magnetic core of the transformer. Eddy current losses generally increase as the square of the frequency, lowering the transformer's efficiency, dissipating additional heat, and reducing its service life.

Negative-sequence harmonics (5th, 11th, 17th, etc.) combine 120 degrees out of phase, similarly to the fundamental harmonic but in a reversed sequence. In generators and motors, these currents produce magnetic fields which oppose the rotation of the shaft and sometimes result in damaging mechanical vibrations.

Switched-mode Power Supplies

A particularly important class of non-linear loads is the millions of personal computers that typically incorporate switched-mode power supplies (SMPS) with rated output power ranging from a few watts to more than 1 kW. Historically, these very-low-cost power supplies incorporated a simple full-wave rectifier that conducted only when the mains instantaneous voltage exceeded the voltage on the input capacitors. This leads to very high ratios of peak-to-average input current, which also lead to a low distortion power factor and potentially serious phase and neutral loading concerns.

A typical switched-mode power supply first converts the AC mains to a DC bus by means of a bridge rectifier or a similar circuit. The output voltage is then derived from this DC bus. The problem with this is that the rectifier is a non-linear device, so the input current is highly non-linear. That means that the input current has energy at harmonics of the frequency of the voltage.

This presents a particular problem for the power companies, because they cannot compensate for the harmonic current by adding simple capacitors or inductors, as they could for the reactive power drawn by a linear load. Many jurisdictions are beginning to legally require power factor correction for all power supplies above a certain power level.

Regulatory agencies such as the EU have set harmonic limits as a method of improving power factor. Declining component cost has hastened implementation of two different methods. To comply with current EU standard EN61000-3-2, all switched-mode power supplies with output power more than 75 W must include passive power factor correction, at least. 80 Plus power supply certification requires a power factor of 0.9 or more.

Power Factor Correction (PFC) in Non-linear Loads

Passive PFC

The simplest way to control the harmonic current is to use a filter that passes current only at line frequency (50 or 60 Hz). The filter consists of capacitors or inductors, and makes a non-linear device look more like a linear load. An example of passive PFC is a valley-fill circuit.

A disadvantage of passive PFC is that it requires larger inductors or capacitors than an equivalent power active PFC circuit. Also, in practice, passive PFC is often less effective at improving the power factor.

Active PFC

610W Continuous @ 40C (670W Peak)
Up to 90% (10dB) Less Noise per Watt
EPS12V / NVIDIA® SLI™ Certified
High Efficiency (83%); .99 Active PFC
+12VDC @ 49A (Large Single Rail)
24-pin, 8-pin, 4-pin M/B Connectors
2 PCI-E and 15 Drive Connectors
Automatic Fan Speed Control Circuit
Black Finish (Copper on request)
5-Year Warranty and Tech Support

Specifications taken from the packaging of a 610 W PC
power supply showing active PFC rating.

Active PFC is the use of power electronics to change the waveform of current drawn by a load to improve the power factor. Some types of the active PFC are buck, boost, buck-boost and synchronous condenser. Active power factor correction can be single-stage or multi-stage.

In the case of a switched-mode power supply, a boost converter is inserted between the bridge rectifier and the main input capacitors. The boost converter attempts to maintain a constant DC bus voltage on its output while drawing a current that is always in phase with and at the same frequency as the line voltage. Another switched-mode converter inside the power supply produces the desired output voltage from the DC bus. This approach requires additional semiconductor switches and control electronics, but permits cheaper and smaller passive components. It is frequently used in practice.

For a three-phase SMPS, the Vienna rectifier configuration may be used to substantially improve the power factor.

SMPSs with passive PFC can achieve power factor of about 0.7–0.75, SMPSs with active PFC, up to

0.99 power factor, while a SMPS without any power factor correction have a power factor of only about 0.55–0.65.

Due to their very wide input voltage range, many power supplies with active PFC can automatically adjust to operate on AC power from about 100 V (Japan) to 240 V (Europe). That feature is particularly welcome in power supplies for laptops.

Dynamic PFC

Dynamic power factor correction (DPFC), sometimes referred to as "real-time power factor correction," is used for electrical stabilization in cases of rapid load changes (e.g. at large manufacturing sites). DPFC is useful when standard power factor correction would cause over or under correction. DPFC uses semiconductor switches, typically thyristors, to quickly connect and disconnect capacitors or inductors to improve power factor.

Power Factor Improvement

Power factor improvement aims at optimal utilization of electrical power, reduction of electricity bills and reduction of power loss.

- Power transformers are independent of power factor. If the power factor is close to unity, for the same KVA rating of the transformer more load can be connected. (Better the power factor lesser will be the current flow).

- Penalties imposed by the power utility companies for non-maintenance of optimal power factor can be avoided.

- Optimal sizing of power cables is possible if the power factor. Low power factor results in higher copper loss (I^2R) loss also more voltage shall be dropped across the cable.

Methods for Power Factor Improvement

The following devices and equipment are used for Power Factor Improvement.

- Static Capacitor
- Synchronous Condenser
- Phase Advancer.

Static Capacitor

We know that most of the industries and power system loads are inductive that take lagging current which decrease the system power factor. For Power factor improvement purpose, Static capacitors are connected in parallel with those devices which work on low power factor.

These static capacitors provide leading current which neutralize (totally or approximately) the lagging inductive component of load current (i.e. leading component neutralize or eliminate the lagging component of load current) thus power factor of the load circuit is improved.

These capacitors are installed in Vicinity of large inductive load e.g Induction motors and transformers etc, and improve the load circuit power factor to improve the system or devises efficiency.

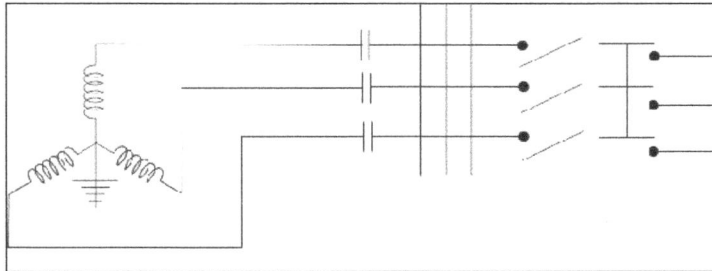

Static Capacitor

Suppose, here is a single phase inductive load which is taking lagging current (I) and the load power factor is $Cos\theta$ as shown in figure.

In figure, above a Capacitor (C) has been connected in parallel with load. Now a current (Ic) is flowing through Capacitor which lead 90° from the supply voltage (note that Capacitor provides leading Current i.e., In a pure capacitive circuit, Current leading 90° from the supply Voltage, in other words, Voltage are 90° lagging from Current). The load current is (I). The Vectors combination of (I) and (Ic) is (I') which is lagging from voltage at θ_2 as shown in figure.

It can be seen from fig that angle of $\theta_2 < \theta_1$ i.e. angle of θ_2 is less than from angle of θ_2. Therefore $Cos\theta_2$ is less than from $Cos\theta_1$ ($Cos\theta_2 > Cos\theta_1$). Hence the load power factor is improved by capacitor.

Also note that after the power factor improvement, the circuit current would be less than from the low power factor circuit current. Also, before and after the power factor improvement, the active component of current would be same in that circuit because capacitor eliminates only the re-active component of current. Also, the Active power (in Watts) would be same after and before power factor improvement.

Advantages

- Capacitor bank offers several advantages over other methods of power factor improvement.

- Losses are low in static capacitors.

- There is no moving part, therefore need low maintenance.

- It can work in normal conditions (i.e. ordinary atmospheric conditions).

- Do not require a foundation for installation.

- They are lightweight so it is can be easy to installed.

Disadvantages

- The age of static capacitor bank is less (8 – 10 years).

- With changing load, we have to ON or OFF the capacitor bank, which causes switching surges on the system.

- If the rated voltage increases, then it causes damage it.

- Once the capacitors spoiled, then repairing is costly.

Synchronous Condenser

When a Synchronous motor operates at No-Load and over-exited then it's called a synchronous Condenser. Whenever a Synchronous motor is over-exited then it provides leading current and works like a capacitor.

When a synchronous condenser is connected across supply voltage (in parallel) then it draws leading current and partially eliminates the re-active component and this way, power factor is improved. Generally, synchronous condenser is used to improve the power factor in large industries.

Advantages

- Long life (almost 25 years).

- High Reliability.

- Step-less adjustment of power factor.

- No generation of harmonics of maintenance.

- The faults can be removed easily.

- It's not affected by harmonics.

- Require Low maintenance (only periodic bearing greasing is necessary).

Disadvantages

- It is expensive (maintenance cost is also high) and therefore mostly used by large power users.

- An auxiliary device has to be used for this operation because synchronous motor has no self-starting torque.

- It produces noise.

Phase Advancer

Phase advancer is a simple AC exciter which is connected on the main shaft of the motor and operates with the motor's rotor circuit for power factor improvement. Phase advancer is used to improve the power factor of induction motor in industries.

As the stator windings of induction motor takes lagging current 90° out of phase with Voltage, therefore the power factor of induction motor is low. If the exciting ampere-turns are excited by external AC source, then there would be no effect of exciting current on stator windings. Therefore the power factor of induction motor will be improved. This process is done by Phase advancer.

Advantages

- Lagging kVAR (Reactive component of Power or reactive power) drawn by the motor is sufficiently reduced because the exciting ampere turns are supplied at slip frequency (fs).

- The phase advancer can be easily used where the use of synchronous motors is Unacceptable.

Disadvantage

- Using Phase advancer is not economical for motors below 200 H.P. (about 150kW).

References

- Electrical-components-used-in-electrical-projects: watelectrical.com, Retrieved 9 August, 2019

- What-is-a-resistor: resistorguide.com, Retrieved 28 January, 2019

- Electrical-resistance: circuitglobe.com, Retrieved 8 April, 2019

- Resistance, electric: rapidtables.com, Retrieved 7 February, 2019

- Electrical-resistance-and-laws-of-resistance: electrical4u.com, Retrieved 13 May, 2019

- Diode-working-principle-and-types-of-diode: electrical4u.com, Retrieved 11 July, 2019

- Transistors-basics-types-baising-modes: elprocus.com, Retrieved 19 March, 2019

- Integrated-circuit, technology: britannica.com, Retrieved 21 February, 2019

- EngineeringFundamentals: maplesoft.com, Retrieved 25 June, 2019

- Basics-of-network-theorems-in-electrical-engineering: elprocus.com, Retrieved 5 March, 2019

- Dc-circuit: circuitglobe.com, Retrieved 11 August, 2019

- What-is-an-ac-circuit: circuitglobe.com, Retrieved 21 June, 2019

- Power-Factor, electric: rapidtables.com, Retrieved 1 April, 2019

- Fink, Donald G.; Beaty, H. Wayne (1978), Standard Handbook for Electrical Engineers (11 ed.), New York: McGraw-Hill, p. 3-29 paragraph 80, ISBN 978-0-07-020974-9

- Power-factor: electricalclassroom.com, Retrieved 20 May, 2019

- Power-factor-improvement-methods-with-their-advantages-disadvantages: electricaltechnology.org, Retrieved 11 January, 2019

4

Magnetic Circuits

The circuits which comprise of one or more closed loop paths having a magnetic flux are known as magnetic circuits. There are a number of phenomena which are studied in relation to these circuits such as hysteresis and inductance. The diverse aspects of magnetic circuits and these connected phenomena have been thoroughly discussed in this chapter.

Magnetic Field

The magnetic field is a field, produced by electric charges in motion. It is a field of force causing a force on material like iron when placed in the vicinity of the field. Magnetic field does not require any medium to propagate; it can propagate even in a vacuum. Also, the energy storing capacity of the magnetic field is greater than the electric field, this distinguishes magnetic field from the electric field, and therefore it is utilised in almost every electromechanical devices like transformers, motors and generators. Earth also has its natural magnetism which protects it from solar waves from the sun. Further, it provides an operating field for a magnetic compass to operate.

Permanent magnets have their own magnetism, and they are made up of ferromagnetic material like iron or nickel or alnico alloys, while electromagnets are coils which produce the magnetic field when an electric current passes through the coil.

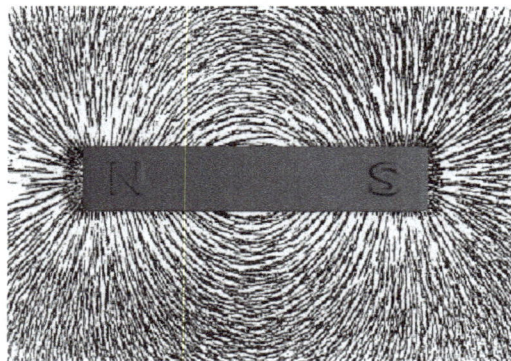

For example, a current carrying conductor produces a magnetic field around the conductor, whose direction is determined by Right-Hand Screw Rule and the strength of field can be varied in accordance with the amount of current flowing in the conductor around the coil. Electromagnets are utilized in various industries for various production and manufacturing processes. The magnetic field has both North pole and a South pole. Monopole does not exist for a magnetic field, unlike electric field where a charge can be isolated. The field line forms a closed loop, as it emanates from North and terminates to South outside a magnet and from South Pole to North Pole inside a magnet.

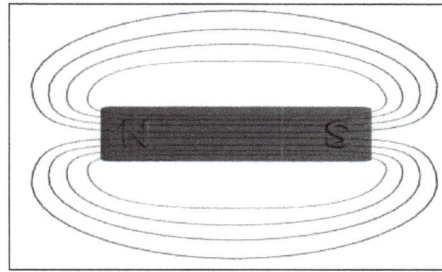

At any point on the field, it has both magnitude and direction, so it is represented by a vector. Magnetic field finds its application in almost every electromechanical device like electric motors and generators. When a current carrying coil is placed in a magnetic field, it experiences a torque. This principle of operation is utilised in electric motor where magnetic torque is produced which exerts a rotating torque on the rotor while in case of generators magnetic field provides a medium for energy exchange between stator and rotor via induction principle. In case of a 3 phase motor, a rotating magnetic field is produced by the 3 phase windings displaced 120 degrees in space. A rotating magnetic field rotates with synchronous speed in the air gap of machines which is required for synchronous motor and induction motors to operate. In order to provide a magnetic medium, machine draws magnetizing current which degrades the power factor of the system. Poor power factor increases the burden on the power system components like transformer and generators, but it is an equally essential component for almost every electromechanical device to operate.

Magnetic Field Lines

A magnetic field line or lines of forces shows the strength of a magnet and the direction of a magnet's force. It was discovered by Michael Faraday to visualize magnetic field.

Magnetic field lines are directed from South Pole to North Pole inside the magnet and from North Pole to South Pole outside the magnet.

A straight current carrying conductor has a magnetic field in the shape of concentric circles around tithe magnetic field of a straight current carrying conductor can be visualised by magnetic field lines.

The direction of a magnetic field produced due to a current carrying conductor rely upon the same direction in which the current is flowing.

The direction of electric field gets reversed if the direction of electric current changes.

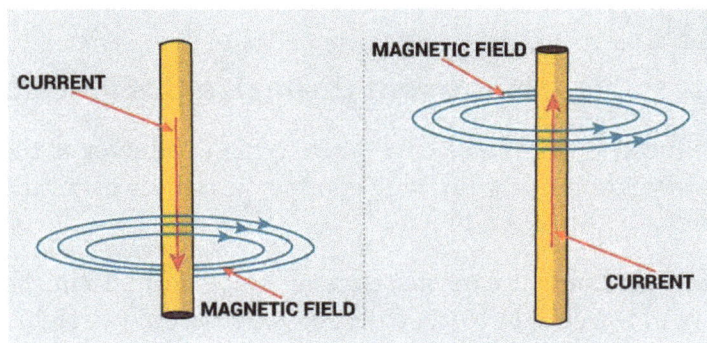

Suppose a straight current carrying conductor is hung vertically, and electric current is flowing from north to south i.e from up to down. In this situation, the direction of magnetic field will be clockwise. And if the same current is flowing from south to north through the same conductor, the direction of magnetic field will be anti-clockwise.

The direction of magnetic field in an electric current through a straight conductor can be represented by using Right-Hand Thumb Rule.

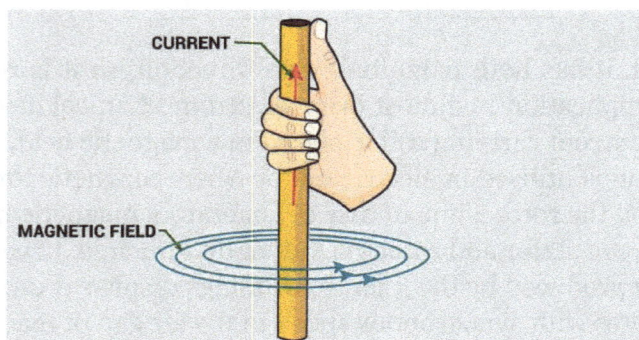

Right-hand Thumb Rule

Assume that you are holding a straight current carrying conductor in your right hand such that the thumb points towards the direction of current. Then your fingers will wrap around the conductor in the direction of field lines of the magnetic field.

Right Hand Thumb rule is also known as Maxwell's corkscrew rule. If we consider ourselves driving a corkscrew in the direction of the current, then the direction of the corkscrew is in the direction of the magnetic field.

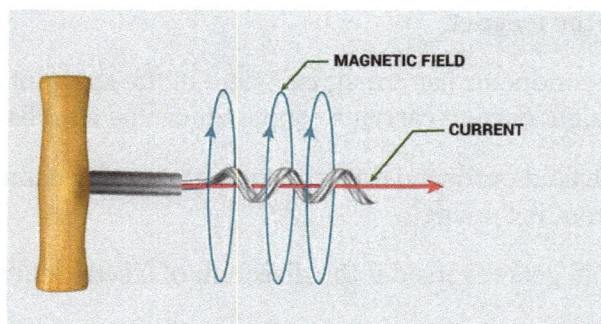

Magnetic Field Due to Flow of Current through a Circular Loop

The magnetic field produced in a circular current carrying conductor is the same as that of the magnetic field due to a straight current carrying conductor and the current carrying circular loop will behave like a magnet.

The magnetic field lines in a current carrying circular loop would be in the shape of concentric circles, and at the centre of the circular wire, field lines will become straight and perpendicular to the plane of coil.

The direction of magnetic field in a circular loop can be recognized using Right Hand Thumb Rule.

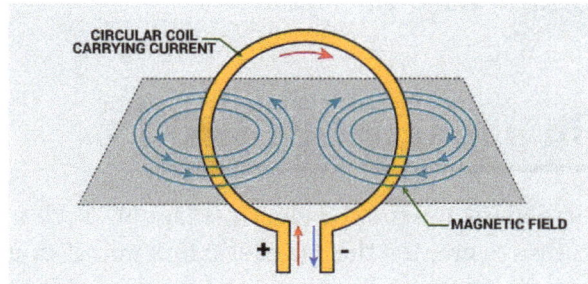

Magnetic Field due to Flow of Current in a Solenoid

A solenoid is a tightly wound helical coil of wire whose diameter is small compared to its length.

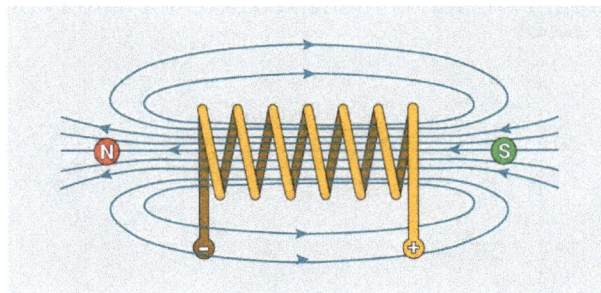

Solenoid

Magnetic field produced by the current carrying solenoid is similar to a bar magnet. One end of a solenoid behaves as a south pole and the other end behaves as a north pole. The magnetic field produced inside a solenoid are parallel which is similar to a bar magnet.

The strong magnetic force produced by a solenoid can be used to magnetise a piece of magnetic material. The magnet so formed is known as an electromagnet.

Direct Current

- Direct Current is the unidirectional flow of electric current. The flow of current in direct current does not change periodically. In case of a direct current, the current flows in a single direction at a steady voltage.

- Direct current power is widely used in low voltage applications such as charging batteries, light aircraft electrical systems.

- By using a rectifier a direct current can be obtained from an alternating current. A rectifier contains electronic elements or electromechanical elements that allow current to flow only in one direction.

- Direct current can also be converted into alternating current by using motor generator set or by an inverter.

- The direction of magnetic field in an electric current through a straight conductor can be represented by using Right Hand Thumb Rule.

Energy Stored in a Magnetic Field

Magnetic field can be of permanent magnet or electro-magnet. Both magnetic fields store some energy. Permanent magnet always creates the magnetic flux and it does not vary upon the other external factors. But electromagnet creates its variable magnetic fields based on how much current it carries. The dimension of this electro-magnet is responsible to create the strength the magnetic field and hence the energy stored in this electromagnet.

First we consider the magnetic field is due to electromagnet i.e. a coil of several no. turns. This coil or inductor is carrying current I when it is connected across a battery or voltage source through a switch.

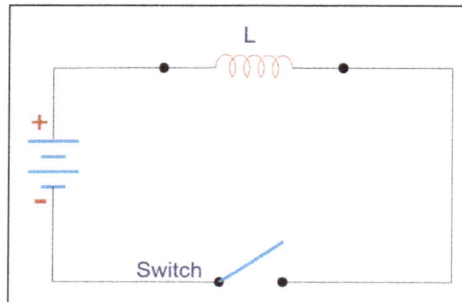

Suppose battery voltage is V volts, value of inductor is L Henry, and current I will flow at steady state.

When the switch is ON, a current will flow from zero to its steady value. But due to self induction a induced voltage appears which is:

$$E = -L\frac{dI}{dt}$$

this E always in the opposite direction of the rate of change of current.

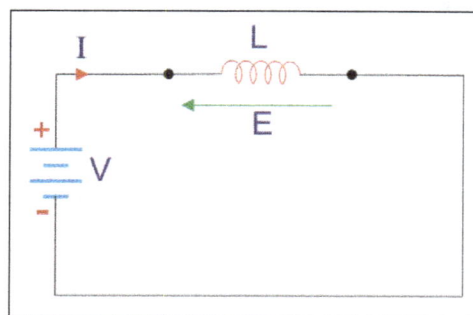

Now here the energy or work done due to this current passing through this inductor is U.

As the current starts from its zero value and flowing against the induced emf E, the energy will grow up gradually from zero value to U.

$$dU = W.dt,$$

where, W is the small power and $W = -E.I$.

So, the energy stored in the inductor is given by,

$$dU = W.dt = -E.Idt = L\frac{dI}{dt}.Idt = LIdI$$

Now integrate the energy from 0 to its final value,

$$U = \int_0^U dU = \int_0^I LIdI = \frac{1}{2}LI^2$$

Again,

$$L = \frac{\mu_0 N^2 A}{l}$$

as per dimension of the coil, where N is the number of turns of the coil, A is the effective cross-sectional area of the coil and l is the effective length of the coil.

Again,

$$I = \frac{H.l}{N}$$

Where, H is the magnetizing force, N is the number of turns of the coil and l is the effective length of the coil:

$$I = \frac{B.l}{\mu_0.N}$$

Now putting expression of L and I in equation of U, we get new expression i.e,

$$U = \frac{\frac{\mu_0 N^2 A}{l}.\frac{B.l}{\mu_0.N}}{2} = \frac{B^2 Al}{2\mu_0}$$

So, the stored energy in a electromagnetic field i.e. a conductor can be calculated from its dimension and flux density.

Energy Stored in the Magnetic Field Due to Permanent Magnet.

Total flux flowing through the magnet cross-sectional area A is φ.

Then, we can write that $\varphi = B.A$, where B is the flux density.

Now this flux φ is of two types, (a) φ_r this is remanent flux of the magnet and (b) φ_d this is demagnetizing flux.

So,

$$\varphi = \varphi_r + \varphi_d$$

as per conservation of the magnetic flux Law:

$$\Phi = A.B_r + A.B_d$$

Again, $B_d = \mu. H$, here H is the magnetic flux intensity.

Now MMF or Magneto Motive Force can be calculated from H and dimension of the magnet,

$$\Delta MMF = H.l$$

where, l is the effective distance between two poles,

$$\Delta MMF = \frac{Bd.l}{\mu} = \frac{l}{\mu}.\varphi d$$

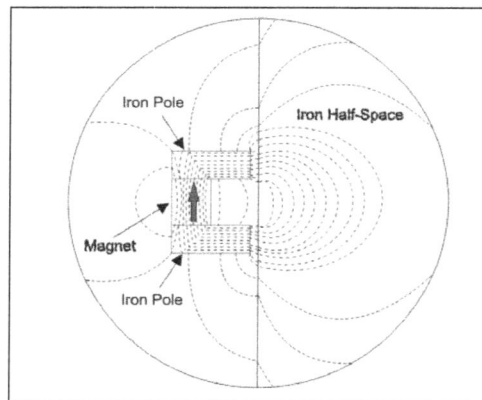

Now to calculate energy we have to first go for the reluctance of the magnetic flux path. Magnet's internal reluctance path that is for demagnetizing is denoted as R_m,

And,

$$R_m = \frac{l}{\mu A}$$

Now W_m, is the energy stored in the magnet's internal reluctance:

$$W_m = \frac{1}{2} \cdot R_m \cdot \varphi \frac{2}{d} = \frac{1}{2} \cdot \frac{l}{\mu.A} \cdot (\mu.H.A) = A.l.\left(\frac{1}{2} \mu H^2 \right)$$

Now, energy density:

$$\frac{W_m}{A.l} = \frac{1}{2}\mu H^2 = W_m'$$

In the figure, the dotted lined box is the magnet and one reluctance path R_l for the mechanical load is connected across the magnet.

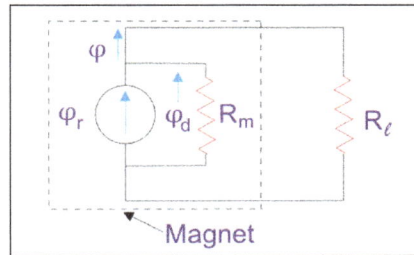

Now apply node equation and loop equation, we get:

$$R_m \cdot \varphi_d + \varphi \cdot R_l = 0,$$

$$\varphi = \varphi_r + \varphi_d$$

$$Again \ \ \varphi = \left(\frac{R_m}{R_m + R_l}\right)\varphi_r \ \ and \ \ \varphi_d = -\varphi_r\left(\frac{R_l}{R_m + R_l}\right)$$

Now, If we do any mechanical work inside a magnetic field, then the energy required W:

$$W = \frac{1}{2}\cdot\left(R_m \cdot \varphi_d^2 + R_l \cdot \varphi^2\right) = \frac{1}{2}\left(\frac{R_l + R_m}{R_l + R_m}\right)\varphi_r^2$$

Again, if we place a electromagnetic coil in the vicinity of a permanent magnet, then this coil will experience a force. To move this coil some work is done. This energy density is the co-energy with respect to the permanent magnet and the coil magnet. Magnetizing flux intensity for the permanent magnet is H and for the coil is H_c.

This co-energy is denoted as,

$$W' = \frac{1}{2}\mu\left(H + H_c\right)^2 = \frac{1}{2}\mu B^2$$

Where, B is the flux density at the coil position near the permanent magnet.

Magnetic Circuit

The closed path followed by magnetic lines of forces is called magnetic circuit. In the magnetic circuit, magnetic flux or magnetic line of forces starts from a point and ends at the same point after completing its path. Flux is generated by magnets; it can be permanent magnet or electro-magnets.

A magnetic circuit is made up of magnetic materials having high permeability such as iron, soft steel, etc. Magnetic circuits are used in various devices like the electric motor, transformers, relays, generators galvanometer, etc. Consider a solenoid was having N turns wound on an iron core. The magnetic flux of ø Weber sets up in the core when the current of I ampere is passed through a solenoid.

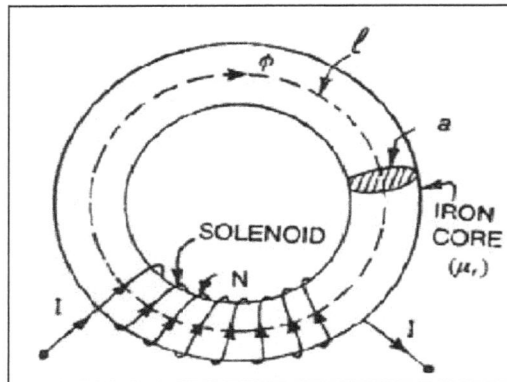

Let,

l = mean length of the magnetic circuit.

A = cross-sectional area of the core.

μr = relative permeability of the core.

Now the flux density in core material,

$$B = \frac{\varphi}{a} \left(Weber / m^2 \right)$$

Magnetising force in the core,

$$H = B / \mu_0 r_0$$

$$H = \varphi / a\, \mu_0 \mu_r \ AT / m \ \left(Ampere\ turns / meter \right)$$

According to work law, the work done in moving a unit pole once round the magnetic circuit is equal to the ampere turns enclosed by the magnetic circuit:

$$Hl = NI$$

$$H = \frac{\varphi}{a\mu_0\mu_r} \times l$$

$$H = NI$$

$$\varphi = \frac{NI}{l / _{a\mu_0\mu_r}}$$

The above equation explains the following points:

1. Directly proportional to a number of turns (N) and current (I). It shows that the flux increase if the number of turns or current increases and decreases when either of the two quantity decreases. NI is the magnetomotive force (MMF).

2. Inversely proportional to $l/a\,\mu_0\mu_r$ where $(l/a\,\mu_0\mu_r)$ is known as reluctance. The lower will be the reluctance the higher will be the flux and vice- verse.

Magnetic Materials

All types of materials and substances possess some kind of magnetic properties. But normally the word "magnetic materials" is used only for ferromagnetic materials; however, materials can be classified into following categories based on the magnetic properties shown by them:

Paramagnetic Materials

The materials which are not strongly attracted to a magnet are known as paramagnetic material. For example: aluminium, tin magnesium etc. Their relative permeability is small but positive. For example: the permeability of aluminium is: 1.00000065. Such materials are magnetized only when placed on a super strong magnetic field and act in the direction of the magnetic field.

Paramagnetic materials have individual atomic dipoles oriented in a random fashion as shown:

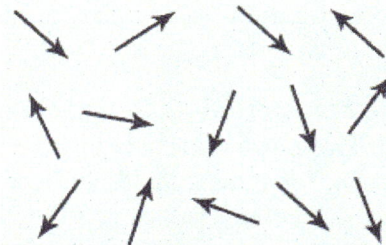

Magnetic Domains in Paramagnetic Materials.

The resultant magnetic force is therefore zero. When a strong external magnetic field is applied, the permanent magnetic dipoles orient them self-parallel to the applied magnetic field and give rise to a positive magnetization. Since, the orientation of the dipoles parallel to the applied magnetic field is not complete, the magnetization is very small.

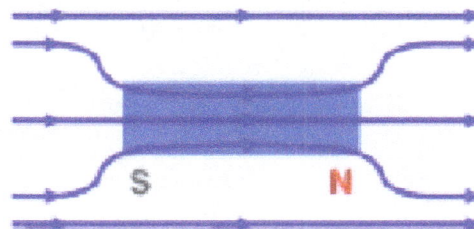

Magnetic field through a paramagnetic material.

Diamagnetic Materials

The materials which are repelled by a magnet such as zinc. Mercury, lead, sulfur, copper, silver, bismuth, wood etc., are known as diamagnetic materials. Their permeability is slightly less than one. For example the relative permeability of bismuth is 0.00083, copper is 0.000005 and wood is 0.9999995. They are slightly magnetized when placed in a very strong magnetic field and act in the direction opposite to that of applied magnetic field.

In diamagnetic materials, the two relatively weak magnetic fields caused due to the orbital revolution and and axial rotation of electrons around nucleus are in opposite directions and cancel each other. Permanent magnetic dipoles are absent in them, Diamegnetic materials have very little to no applications in electrical engineering.

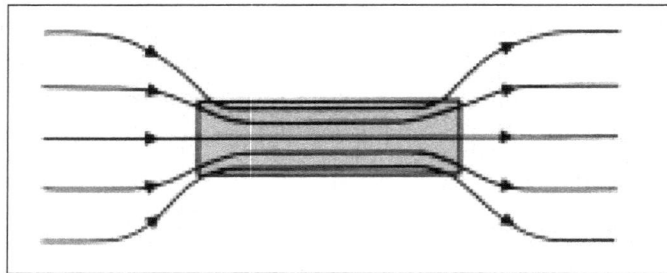

Magnetic field through a diamagnetic material.

Ferromagnetic Materials

The materials which are strongly attracted by a magnetic field or magnet is known as ferromagnetic material for e.g.: iron, steel, nickel, cobalt etc. The permeability off these materials is very very high (ranging up to several hundred or thousand).

The opposite magnetic effects of electron orbital motion and electron spin do not eliminate each other in an atom of such a material. There is a relatively large contribution from each atom which aids in the establishment of an internal magnetic field, so that when the material is placed in a magnetic field, it's value is increased many times the value that was present in the free space before the material was placed there.

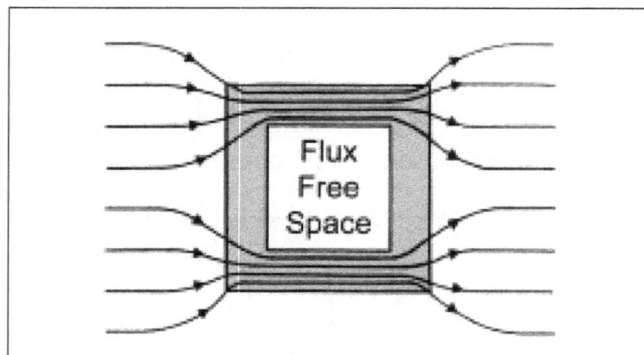

Flux
Free
Space

Magnetic field through a ferromagnetic material.

For the purpose of electrical engineering it will suffice to classify the materials as simply ferromagnetic and and non-ferromagnetic materials. The latter includes material of relative permeability practically equal to unity while the former have relative permeability many times greater than unity. Paramagnetic and diamagnetic material falls in the non-ferromagnetic materials.

Ferromagnetic materials can be further classified into two types which are:

1. Soft Ferromagnetic materials:

They have high relative permeability, low coercive force, easily magnetized and demagnetized and have extremely small hysteresis. Soft ferromagnetic materials are iron and it's various alloys with materials like nickel, cobalt, tungsten and aluminium. Ease of magnetization and demagnetization makes them highly suitable for applications involving changing magnetic flux as in electromagnets, electric motors, generators, transformers, inductors, telephone receivers, relays etc. They are also useful for magnetic screening. Their properties may be greatly enhanced through careful manufacturing and and by heating and slow annealing so as to achieve a high degree of crystal purity. Large magnetic moment at room temperate makes soft ferromagnetic materials extremely useful for magnetic circuits but ferromagnetics are very good conductors and suffer energy loss from eddy current produced within them. There is additional energy loss due to the fact that magnetization does not proceed smoothly but in minute jumps. This loss is called magnetic residual loss and it depends purely on the frequency of the changing flux density and not on its magnitude.

2. Hard Ferromagnetic materials:

They have relatively low permeability, and very high coercive force. These are difficult o magnetize and demagnetize. Typical hard ferromagnetic materials include cobalt steel and various ferromagnetic alloys of cobalt, aluminium and nickel. They retain high percentage of their magnetization and have relatively high hysteresis loss. They are highly suited for use as permanent magnet as speakers, measuring instruments etc.

Ferrites

Ferrites are a special group of ferromagnetic materials that occupy an intermediate position between ferromagnetic and non-ferromagnetic materials. They consist of extremely fine particles of a ferromagnetic material possessing high permeability, and are held together with a binding resin. The magnetization produced in ferrites is large enough to be of commercial value but their magnetic saturation are not as high as those of ferromagnetic materials. As in the case of ferromagnetics, ferrites may be soft or hard ferrites.

1. Soft Ferrites:

Ceramic magnets also called ferromagnetic ceramics are made of an iron oxide, Fe_2O_3, with one or more divalent oxide such as NiO, MnO or ZnO. These magnets have a square hysteresis loop and high resistance and demagnetization are valued for magnets for computing machines where a high resistance is desired. The great advantage of ferrites is their high resistivity. Commercial magnets have resistivity as high as 10^9 ohm-cm. Eddy currents resulting from an alternating fields are therefore, reduced to minimum, and the range of application of these magnetic materials is extended to high frequencies, even to microwaves. Ferrites are carefully made by mixing powdered oxides, compacting and sintering at high temperature. High frequency transformers in televisions and frequency modulated receivers are almost always made with ferrite cores.

2. Hard Ferrites:

These are ceramic permanent magnetic materials. The most important family of hard ferrites has the basic composition of $MO.Fe_2O_3$ where M is barium (Ba) ion or strontium (Sr) ion. These materials have a hexagonal structure and low in cost and density. Hard ferrites are used in generators, relays and motors. Electronic applications include magnets for loud speakers, telephone ringers and receivers. They are also used in holding devices for door closer, seals, latches and in several toy designs.

Electromagnetic Induction

The induction of an electromotive force by the motion of a conductor across a magnetic field or by a change in magnetic flux in a magnetic field is called Electromagnetic Induction. Electromagnetic Induction is the basic principle of operation of transformers, motors and generators.

Electromagnetic Induction was first discovered way back in the 1830's by Michael Faraday. Faraday noticed that when he moved a permanent magnet in and out of a coil or a single loop of wire it induced an ElectroMotive Force or emf, in other words a Voltage, and therefore a current was produced.

So what Michael Faraday discovered was a way of producing an electrical current in a circuit by using only the force of a magnetic field and not batteries. This then lead to a very important law linking electricity with magnetism, Faraday's Law of Electromagnetic Induction.

When the magnet shown is moved "towards" the coil, the pointer or needle of the Galvanometer, which is basically a very sensitive centre zero'ed moving-coil ammeter, will deflect away from its centre position in one direction only. When the magnet stops moving and is held stationary with regards to the coil the needle of the galvanometer returns back to zero as there is no physical movement of the magnetic field.

Likewise, when the magnet is moved "away" from the coil in the other direction, the needle of the galvanometer deflects in the opposite direction with regards to the first indicating a change in polarity. Then by moving the magnet back and forth towards the coil the needle of the galvanometer will deflect left or right, positive or negative, relative to the directional motion of the magnet.

Electromagnetic Induction by a Moving Magnet

Likewise, if the magnet is now held stationary and ONLY the coil is moved towards or away from the magnet the needle of the galvanometer will also deflect in either direction. Then the action of moving a coil or loop of wire through a magnetic field induces a voltage in the coil with the magnitude of this induced voltage being proportional to the speed or velocity of the movement.

Then we can see that the faster the movement of the magnetic field the greater will be the induced emf or voltage in the coil, so for Faraday's law to hold true there must be "relative motion" or movement between the coil and the magnetic field and either the magnetic field, the coil or both can move.

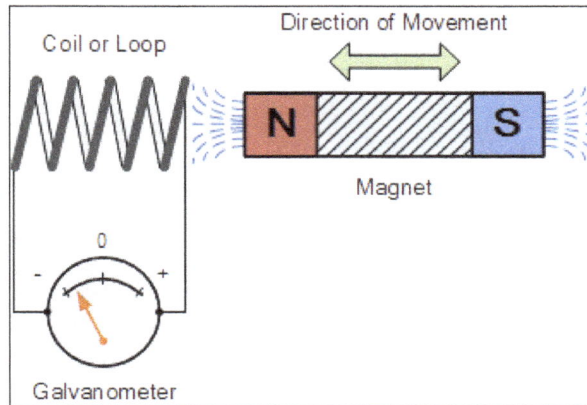

Faraday's Law of Induction

From the above description we can say that a relationship exists between an electrical voltage and a changing magnetic field to which Michael Faraday's famous law of electromagnetic induction states: "that a voltage is induced in a circuit whenever relative motion exists between a conductor and a magnetic field and that the magnitude of this voltage is proportional to the rate of change of the flux".

In other words, Electromagnetic Induction is the process of using magnetic fields to produce voltage, and in a closed circuit, a current.

So how much voltage (emf) can be induced into the coil using just magnetism. Well this is determined by the following 3 different factors.

1. Increasing the number of turns of wire in the coil: By increasing the amount of individual conductors cutting through the magnetic field, the amount of induced emf produced will be the sum of all the individual loops of the coil, so if there are 20 turns in the coil there will be 20 times more induced emf than in one piece of wire.

2. Increasing the speed of the relative motion between the coil and the magnet: If the same coil of wire passed through the same magnetic field but its speed or velocity is increased, the wire will cut the lines of flux at a faster rate so more induced emf would be produced.

3. Increasing the strength of the magnetic field: If the same coil of wire is moved at the same speed through a stronger magnetic field, there will be more emf produced because there are more lines of force to cut.

If we were able to move the magnet in the diagram above in and out of the coil at a constant speed and distance without stopping we would generate a continuously induced voltage that would alternate between one positive polarity and a negative polarity producing an alternating or AC output voltage and this is the basic principle of how an electrical generator works similar to those used in dynamos and car alternators.

In small generators such as a bicycle dynamo, a small permanent magnet is rotated by the action of the bicycle wheel inside a fixed coil. Alternatively, an electromagnet powered by a fixed DC voltage can be made to rotate inside a fixed coil, such as in large power generators producing in both cases an alternating current.

Simple Generator using Magnetic Induction

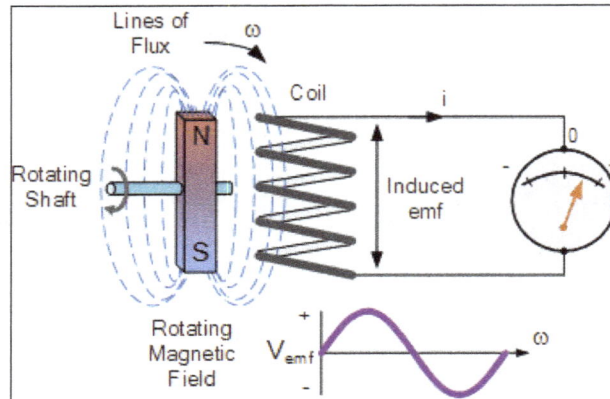

The simple dynamo type generator above consists of a permanent magnet which rotates around a central shaft with a coil of wire placed next to this rotating magnetic field. As the magnet spins, the magnetic field around the top and bottom of the coil constantly changes between a north and a south pole. This rotational movement of the magnetic field results in an alternating emf being induced into the coil as defined by Faraday's law of electromagnetic induction.

The magnitude of the electromagnetic induction is directly proportional to the flux density, β the number of loops giving a total length of the conductor, l in meters and the rate or velocity, v at which the magnetic field changes within the conductor in meters/second or m/s, giving by the motional emf expression:

Faraday's Motional Emf Expression

$$\varepsilon = -\beta.\ell.v \text{ volts}$$

If the conductor does not move at right angles (90°) to the magnetic field then the angle θ° will be added to the above expression giving a reduced output as the angle increases:

$$\varepsilon = -\beta.\ell.v \sin\theta \text{ volts}$$

Lenz's Law of Electromagnetic Induction

Faraday's Law tells us that inducing a voltage into a conductor can be done by either passing it through a magnetic field, or by moving the magnetic field past the conductor and that if this conductor is part of a closed circuit, an electric current will flow. This voltage is called an induced emf as it has been induced into the conductor by a changing magnetic field due to electromagnetic induction with the negative sign in Faraday's law telling us the direction of the induced current (or polarity of the induced emf).

But a changing magnetic flux produces a varying current through the coil which itself will produce its own magnetic field This self-induced emf opposes the change that is causing it and the faster the rate of change of current the greater is the opposing emf. This self-induced emf will, by Lenz's law oppose the change in current in the coil and because of its direction this self-induced emf is generally called a back-emf.

Lenz's Law states that:" the direction of an induced emf is such that it will always oppose the change that is causing it". In other words, an induced current will always oppose the motion or change which started the induced current in the first place and this idea is found in the analysis of Inductance.

Likewise, if the magnetic flux is decreased then the induced emf will oppose this decrease by generating and induced magnetic flux that adds to the original flux. Lenz's law is one of the basic laws in electromagnetic induction for determining the direction of flow of induced currents and is related to the law of conservation of energy.

According to the law of conservation of energy which states that the total amount of energy in the universe will always remain constant as energy cannot be created nor destroyed. Lenz's law is derived from Michael Faraday's law of induction.

One final comment about Lenz's Law regarding electromagnetic induction. We now know that when a relative motion exists between a conductor and a magnetic field, an emf is induced within the conductor.

But the conductor may not actually be part of the coils electrical circuit, but may be the coils iron core or some other metallic part of the system, for example, a transformer. The induced emf within this metallic part of the system causes a circulating current to flow around it and this type of core current is known as an Eddy Current.

Eddy currents generated by electromagnetic induction circulate around the coils core or any connecting metallic components inside the magnetic field because for the magnetic flux they are acting like a single loop of wire. Eddy currents do not contribute anything towards the usefulness of the system but instead they oppose the flow of the induced current by acting like a negative force generating resistive heating and power loss within the core. However, there are electromagnetic induction furnace applications in which only eddy currents are used to heat and melt ferromagnetic metals.

Eddy Currents Circulating in a Transformer

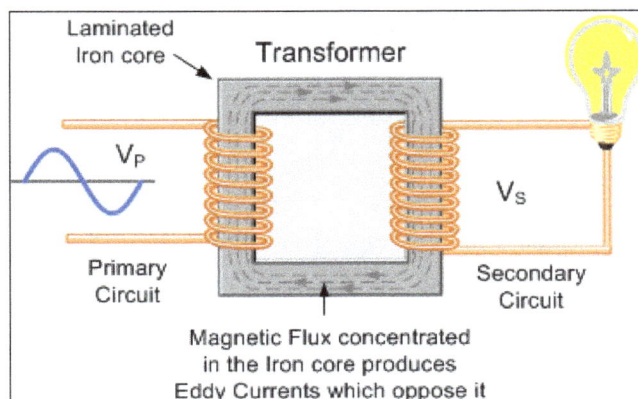

Laminated Iron core ⟍   Transformer

V_P

Primary Circuit

V_S

Secondary Circuit

Magnetic Flux concentrated
in the Iron core produces
Eddy Currents which oppose it

The changing magnetic flux in the iron core of a transformer above will induce an emf, not only in the primary and secondary windings, but also in the iron core. The iron core is a good conductor, so the currents induced in a solid iron core will be large. Furthermore, the eddy currents flow in a direction which, by Lenz's law, acts to weaken the flux created by the primary coil. Consequently, the current in the primary coil required to produce a given B field is increased, so the hysteresis curves are fatter along the H axis:

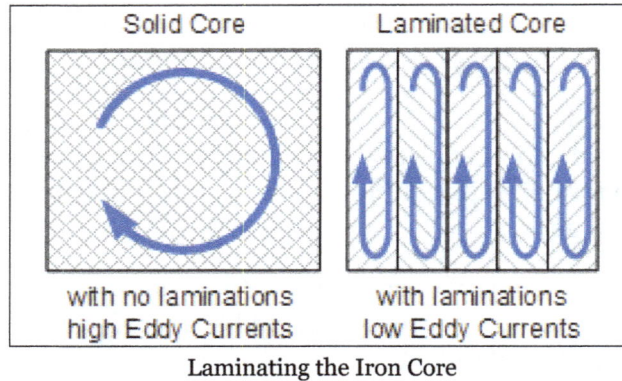

Laminating the Iron Core

Eddy current and hysteresis losses cannot be eliminated completely, but they can be greatly reduced. Instead of having a solid iron core as the magnetic core material of the transformer or coil, the magnetic path is "laminated".

These laminations are very thin strips of insulated (usually with varnish) metal joined together to produce a solid core. The laminations increase the resistance of the iron-core thereby increasing the overall resistance to the flow of the eddy currents, so the induced eddy current power-loss in the core is reduced, and it is for this reason why the magnetic iron circuit of transformers and electrical machines are all laminated.

Applications of Electromagnetic Induction

1. Electromagnetic induction in AC generator

2. Electrical Transformers

3. Magnetic Flow Meter

Electromagnetic Induction in AC Generator

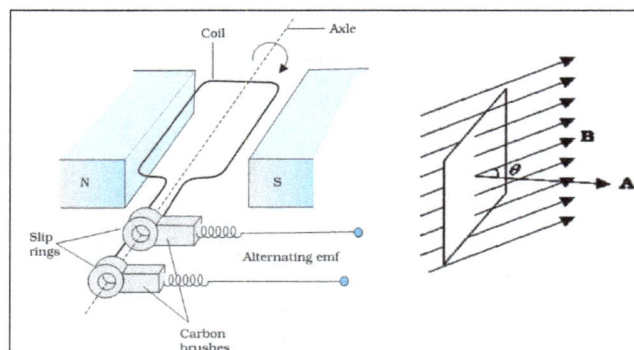

One of the important applications of electromagnetic induction is the generation of alternating current.

The AC generator with an output capacity of 100 MV is a more evolved machine. As the coil rotates in a magnetic field B, the effective area of the loop is A cosθ, where θ is the angle between A and B. This is a method of producing a flux change is the principle of operation of a simple ac generator. The axis of rotation coil is perpendicular to the direction of the magnetic field. The rotation of the coil causes the magnetic flux through it to change, so an emf keeps inducing in the coil.

Electrical Transformers

Another important application of electromagnetic induction is an electrical transformer. A transformer is a device that changes ac electric power at one voltage level to another level through the action of a magnetic field. A step-down transformer is the one in which the voltage is higher in the primary than the secondary voltage. Whereas the one in which the secondary voltage has more turns is a step-up transformer. Power companies use a step transformer to boost the voltage to 100 kV that reduces the current and minimizes the loss of power in transmission lines. On the other end, household circuits use step-down transformers to decrease the voltage to the 120 or 240 V in them.

Inductance

A current generated in a conductor by a changing magnetic field is proportional to the rate of change of the magnetic field. This effect is called INDUCTANCE and is given the symbol L. It is measured in units called the henry (H) named after the American Physicist Joseph Henry (1797-1878). One henry is the amount of inductance required to produce an emf of 1 volt in a conductor when the current in the conductor changes at the rate of 1 Ampere per second. The Henry is a rather large unit for use in electronics, with the milli-henry (mH) and micro-henry (μH) being more common. These units describe one thousandth and one millionth of a henry respectively.

Although the henry is given the symbol (capital)H the name henry, applied to the unit of inductance uses a lower case h. The plural form of the henry may be henries or henrys; the American National Institute of Standards and Technology recommends that in US publications henries is used.

Factors affecting Inductance

The amount of inductance in an inductor is dependent on:

- The number of turns of wire in the inductor.

- The material of the core.

- The shape and size of the core.

- The shape, size and arrangement of the wire making up the coils.

Because inductance (in henries) depends on so many variable quantities, it is quite difficult to calculate accurately; numerous formulae have been developed to take different design features into account. Also these formulae often need to use special constants and tables of conversion data to work with the required degree of accuracy. The use of computer programs and computer-aided design has eased the situation somewhat. However, external effects caused by other components and wiring near the inductor, can also affect its value of inductance once it is assembled in a circuit, so when an accurate value of inductance is required, one approach is to calculate an approximate value, and design the inductor so that it is adjustable.

A typical formula for approximating the inductance value of an inductor is given. This particular version is designed to calculate the inductance of "A solenoid wound with a single layer of turns of infinitely thin tape rather than wire, and with the turns evenly and closely spaced."

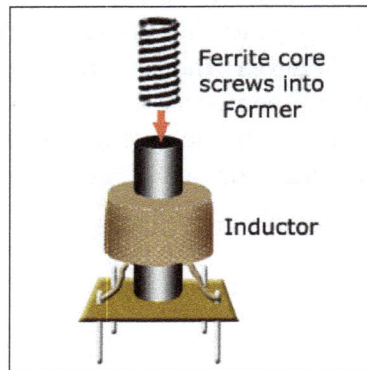

Ferrite core screws into Former

Inductor

A Miniature Variable Inductor.

$$L = \frac{\left(d^2 n^2\right)}{l + 0.45\,d}$$

Where,

- L is the inductance in henries.

- d is the diameter of the coil in metres.

- n is the number of turns in the coil.

- l is the length of the coil in metres.

For coils not conforming exactly to the above specification extra factors must be incorporated. Figure illustrates one way of producing a sufficiently accurate inductance, used in some HF and RF circuits. A miniature inductor is wound on a plastic former, into which a ferrite (iron dust) core is screwed sufficiently to provide a core giving the right amount of inductance.

Voltage and e.m.f.

A voltage induced into a conductor is called an e.m.f. (electro motive force) because its source is the changing magnetic field around and external to the conductor. Any externally produced voltage (including those produced by an external battery or power supply) is called an e.m.f., whilst a voltage (a potential difference or p.d.) across an internal component in a circuit is called a voltage.

Back e.m.f.

A back e.m.f.(also called a Counter e.m.f.) is an emf created across an inductor by the changing magnetic flux around the conductor, produced by a change in current in the inductor. Its value can be calculated using the formula:

$$E = -L\frac{\Delta I}{\Delta t}$$

Where,

- E is the induced back e.m.f. in volts.
- L is the inductance of the coil in henries.
- ΔI is the change in current, in amperes.
- Δt is the time taken for the change in current, in seconds.

Δ denotes a difference or change in a property.

So the formula describes the back emf as depending on the inductance (in henries) multiplied by the rate of change in current (in amperes per second).

The minus sign before L indicates that the polarity of the induced back emf will be reversed compared with the changing voltage across the conductor that originally caused the changing current and consequent changing magnetic field.

Self Inductance

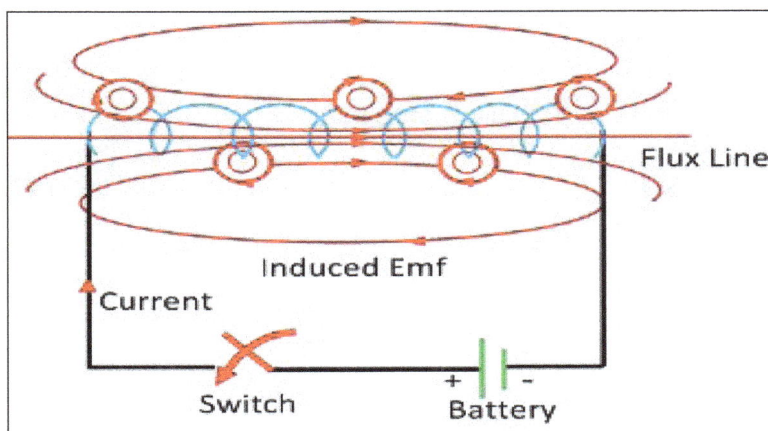

Self Inductance.

Self-inductance of the coil is defined as the property of the coil due to which it opposes the change of current flowing through it. Inductance is attained by a coil due to the self-induced emf produced in the coil itself by changing the current flowing through it.

If the current in the coil is increasing, the self-induced emf produced in the coil will oppose the rise of current that means the direction of the induced emf is opposite to the applied voltage.

If the current in the coil is decreasing, the emf induced in the coil is in such a direction as to oppose the fall of current; this means that the direction of the self-induced emf is same as that of the applied voltage. Self-inductance does not prevent the change of current, but it delays the change of current flowing through it.

This property of the coil only opposes the changing current (alternating current) and does not affect the steady current that is (direct current) when flows through it. The unit of inductance is Henry (H).

Expression for Self Inductance

You can determine the self-inductance of a coil by the following expression:

$$e = L\frac{dI}{dt}$$

or

$$L = \frac{e}{dI / dt}$$

The above expression is used when the magnitude of self-induced emf (e) in the coil and the rate of change of current (dI/dt) is known.

Putting the following values in the above equations as e = 1 V, and dI/dt = 1 A/s then the value of Inductance will be L = 1 H.

Hence, from the above derivation, a statement can be given that a coil is said to have an inductance of 1 Henry if an emf of 1 volt is induced in it when the current flowing through it changes at the rate of 1 Ampere/second.

The expression for Self Inductance can also be given as:

$$e = L\frac{dI}{dt} = \frac{d}{dt}(LI) \; also \; e = N\frac{d\varphi}{dt} = \frac{d}{dt}(N\varphi)$$

$$LI = N\varphi \; or \; L = \frac{N\varphi}{I} \; Henry$$

Where,

      N – Number of turns in the coil.

      Φ – Magnetic flux.

      I – Current flowing through the coil.

The following points can be drawn about Self Inductance

- The value of the inductance will be high if the magnetic flux is stronger for the given value of current.

- The value of the Inductance also depends upon the material of the core and the number of turns in the coil or solenoid.

- The higher will be the value of the inductance in Henry, the rate of change of current will be lower.

- 1 Henry is also equal to 1 Weber/ampere

The solenoid has large self-inductance.

Mutual Inductance

Mutual Inductance is the interaction of one coils magnetic field on another coil as it induces a voltage in the adjacent coil. Mutual Inductance is the basic operating principal of the transformer, motors, generators and any other electrical component that interacts with another magnetic field. Then we can define mutual induction as the current flowing in one coil that induces a voltage in an adjacent coil.

But mutual inductance can also be a bad thing as "stray" or "leakage" inductance from a coil can interfere with the operation of another adjacent component by means of electromagnetic induction, so some form of electrical screening to a ground potential may be required.

The amount of mutual inductance that links one coil to another depends very much on the relative positioning of the two coils. If one coil is positioned next to the other coil so that their physical distance apart is small, then nearly all of the magnetic flux generated by the first coil will interact with the coil turns of the second coil inducing a relatively large emf and therefore producing a large mutual inductance value.

Likewise, if the two coils are farther apart from each other or at different angles, the amount of induced magnetic flux from the first coil into the second will be weaker producing a much smaller induced emf and therefore a much smaller mutual inductance value. So the effect of mutual inductance is very much dependent upon the relative positions or spacing, (S) of the two coils.

Mutual Inductance between Coils

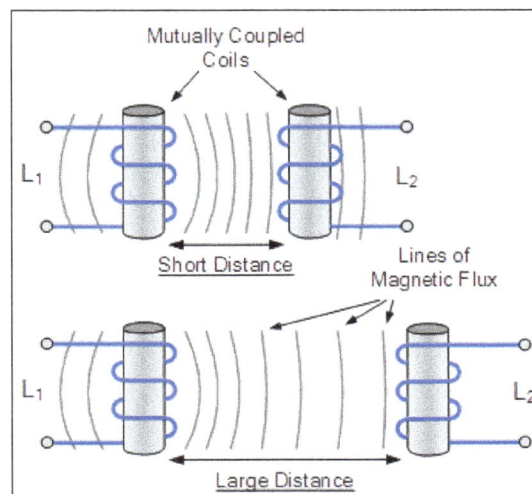

The mutual inductance that exists between the two coils can be greatly increased by positioning them on a common soft iron core or by increasing the number of turns of either coil as would be found in a transformer.

If the two coils are tightly wound one on top of the other over a common soft iron core unity coupling is said to exist between them as any losses due to the leakage of flux will be extremely small. Then assuming a perfect flux linkage between the two coils the mutual inductance that exists between them can be given as,

$$M = \frac{\mu_0 \mu_r N_1 N_2 A}{\ell}$$

Where,

- μ_o is the permeability of free space ($4.\pi.10^{-7}$).

- μ_r is the relative permeability of the soft iron core.

- N is in the number of coil turns.

- A is in the cross-sectional area in m².

- l is the coils length in meters.

Mutual Induction

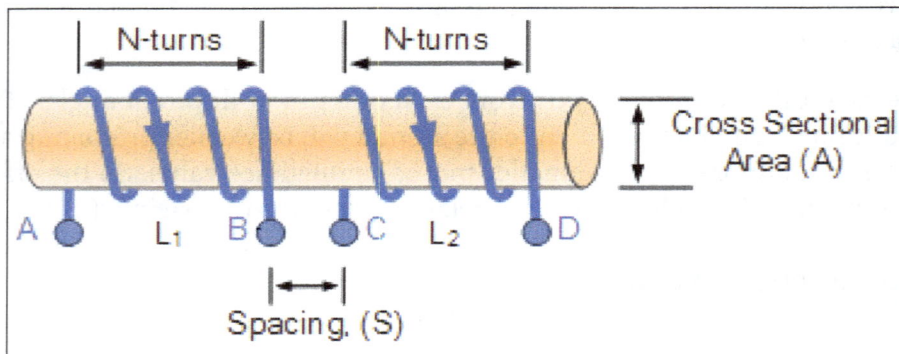

Here the current flowing in coil one, L_1 sets up a magnetic field around itself with some of these magnetic field lines passing through coil two, L_2 is giving us mutual inductance. Coil one has a current of I_1 and N_1 turns while, coil two has N_2 turns. Therefore, the mutual inductance, M_{12} of coil two that exists with respect to coil one depends on their position with respect to each other and is given as:

$$M_{12} = \frac{N_2 \Phi_{12}}{I_1}$$

Likewise, the flux linking coil one, L_1 when a current flows around coil two, L_2 is exactly the same as the flux linking coil two when the same current flows around coil one above, then the mutual inductance of coil one with respect of coil two is defined as M_{21}. This mutual inductance is true irrespective of the size, number of turns, relative position or orientation of the two coils. Because of this, we can write the mutual inductance between the two coils as:

$$M_{12} = M_{21} = M.$$

Then we can see that self inductance characterises an inductor as a single circuit element, while mutual inductance signifies some form of magnetic coupling between two inductors or coils, depending on their distance and arrangement, the self inductance of each individual coil is given as:

$$L_1 = \frac{\mu_0 \mu_r N_1^2 A}{\ell} \quad and \quad L_2 = \frac{\mu_0 \mu_r N_2^2 A}{\ell}$$

By cross-multiplying the two equations above, the mutual inductance, M that exists between the two coils can be expressed in terms of the self inductance of each coil,

$$M^2 = L_1 L_2$$

giving us a final and more common expression for the mutual inductance between the two coils.

Mutual Inductance between Coils

$$M = \sqrt{L_1 L_2} H$$

However, the above equation assumes zero flux leakage and 100% magnetic coupling between the two coils, L_1 and L_2. In reality there will always be some loss due to leakage and position, so the magnetic coupling between the two coils can never reach or exceed 100%, but can become very close to this value in some special inductive coils.

If some of the total magnetic flux links with the two coils, this amount of flux linkage can be defined as a fraction of the total possible flux linkage between the coils. This fractional value is called the coefficient of coupling and is given the letter k.

Coupling Coefficient

Generally, the amount of inductive coupling that exists between the two coils is expressed as a fractional number between 0 and 1 instead of a percentage (%) value, where 0 indicates zero or no inductive coupling, and 1 indicating full or maximum inductive coupling.

In other words, if k = 1 the two coils are perfectly coupled, if k > 0.5 the two coils are said to be tightly coupled and if k < 0.5 the two coils are said to be loosely coupled. Then the equation above which assumes a perfect coupling can be modified to take into account this coefficient of coupling, k and is given as:

Coupling Factor between Coils

$$k = \frac{M}{\sqrt{L_1 L_2}} \quad or \quad M = k\sqrt{L_1 L_2}$$

When the coefficient of coupling, k is equal to 1, (unity) such that all the lines of flux of one coil cuts all of the turns of the second coil, that is the two coils are tightly coupled together, the resulting mutual inductance will be equal to the geometric mean of the two individual inductances of the coils.

Also when the inductances of the two coils are the same and equal, L_1 is equal to L_2, the mutual inductance that exists between the two coils will equal the value of one single coil as the square root of two equal values is the same as one single value as shown:

$$M = \sqrt{L_1 L_2} = L.$$

Hysteresis

The phenomenon of flux density B lagging behind the magnetizing force H in a magnetic material is known as Magnetic Hysteresis. In other words, when the magnetic material is magnetized first in one direction and then in the other direction, completing one cycle of magnetization, it is found that the flux density B lags behind the applied magnetization force H.

There are various types of magnetic materials such as paramagnetic, diamagnetic, ferromagnetic, ferromagnetic and antiferromagnetic materials. Ferromagnetic materials are mainly responsible for the generation of the hysteresis loop.

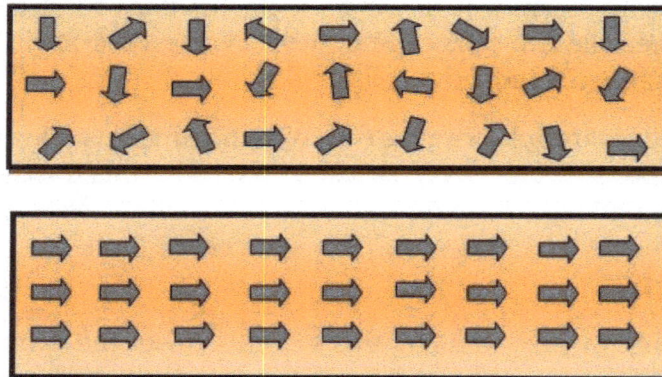

When the magnetic field in not applied the ferromagnetic material behaves like a paramagnetic material. This means that at the initial stage the dipole of the ferromagnetic material is not aligned, they are randomly placed. As soon as the magnetic field is applied to the ferromagnetic material, its dipole moments align themselves in one particular direction as shown in the above figure, resulting in a much stronger magnetic field.

For understanding the phenomenon of the magnetic hysteresis, consider a ring of magnetic material wound uniformly with solenoid. The solenoid is connected to a DC source through a Double pole double throw (D.P.D.T) reversible switch as shown in the figure.

Initially, the switch is in position 1. By decreasing the value of R the value of the current in the solenoid increases gradually resulting in a gradual increase in field intensity H, the flux density also increases till it reaches the saturation point a and the curve obtained is oa. Saturation occurs when on increasing the current the dipole moment or the molecules of the magnet material align itself in one direction.

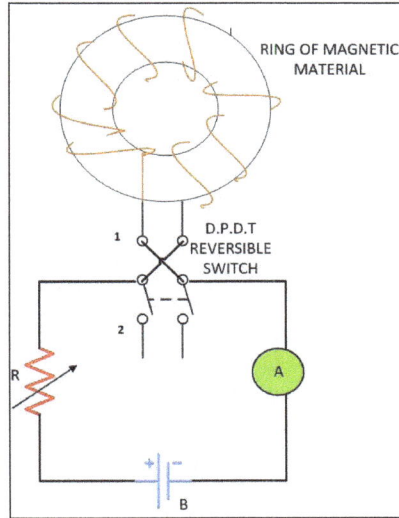

Now by decreasing the current in the solenoid to zero the magnetizing force is gradually reduced to zero, but the value of flux density will not be zero as it still has the value ob when H=o, so the curve obtained is ab as shown in the figure. This value ob of flux density is because of the residual magnetism.

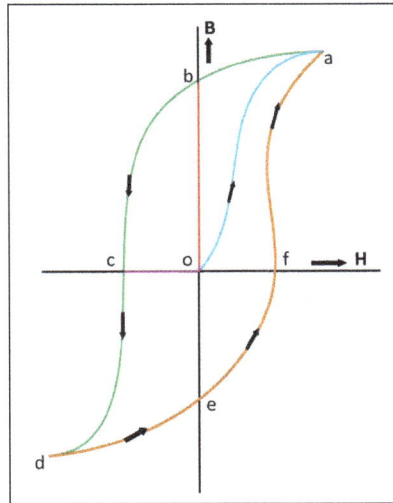

Hysteresis Loop.

Residual Magnetism

The value of the flux density ob retained by the magnetic material is called residual magnetism, and the power of retaining it is known as Retentivity of the material.

Now to demagnetize the magnetic ring, the position of the D.P.D.T reversible switch is changed to position 2 and thus, the direction of flow of the current in the solenoid is reversed resulting in reverse magnetizing force H. When H is increased in reverse direction, the flux density starts decreasing and becomes zero (B=0) and the curve shown above follows the path bc. The residual magnetism of the material is removed by applying the magnetizing force known as Coercive force in the opposite direction.

Coercive Force

The value of the magnetizing force oc required to wipe out the residual magnetism ob is called Coercive force shown by pink color in the hysteresis curve shown above.

Now to complete the hysteresis loop the magnetizing force H is further increased in the reverse direction till it reaches the saturation point d but in the negative direction, the curve traces the path cd. The value of H is reduced to zero H = 0 and the curve obtains the path de, where oe is residual magnetism when the curve is in the negative direction.

The position of the switch is changed to 1 again from the position 2 and the current in the solenoid is again increased as done in the magnetization process and due to this H is increased in the positive direction tracing the path as efa, and finally the hysteresis loop is complete. In the curve again of is the magnetizing force, also known as the Coercive force required to remove the residual magnetism oe.

Here the total Coercive force required to wipe off the residual magnetism in one complete cycle is denoted by cf. It is clear that the flux density B always lags behind the magnetizing force H. Hence the loop 'abcdefa' is called the Magnetic Hysteresis loop or Hysteresis Curve.

Magnetic hysteresis results in the dissipation of wasted energy in the form of heat. The energy wasted is proportional to the area of the magnetic hysteresis loop. Mainly there are two types of magnetic material, soft magnetic material and hard magnetic material.

Soft Magnetic Material

The soft magnetic material has a narrow magnetic hysteresis loop as shown in the figure. Which has a small amount of dissipated energy. They are made up of material like iron, silicon steel, etc.

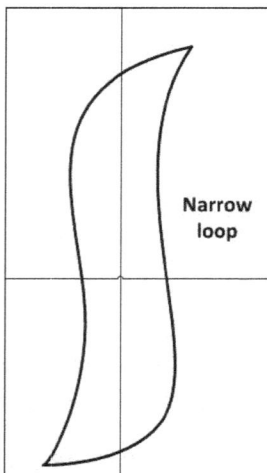

Soft Magnetic Material Loop.

- It is used in the devices that require alternating magnetic field.

- It has low coercivity.

- Low magnetization.

- Low retentivity.

Hard Magnetic Material

The Hard magnetic material has a wider hysteresis loop as shown in the figure and results in a large amount of energy dissipation and the demagnitisation process is more difficult to achieve.

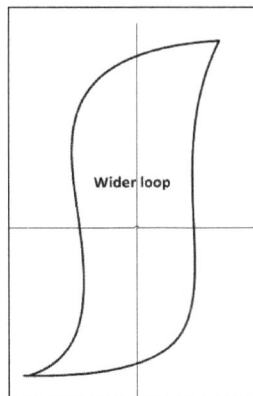

Wider loop

Hard Magnetic Material Loop.

- It has high retentivity.

- High coercivity.

- High saturation.

Applications of Magnetic Hysteresis

- Magnetic material having a wider hysteresis loop is used in the devices like magnetic tape, hard disk, credit cards, and audio recordings as its memory is not easily erased.

- Magnetic materials having a narrow hysteresis loop are used as electromagnets, solenoid, transformers and relays which require minimum energy dissipation.

References

- Magnetic-field: electrical4u.com: electrical4u.com, Retrieved 23 July, 2019

- Energy-stored-in-a-magnetic-field: electrical4u.com, Retrieved 13 February, 2019

- What-is-a-magnetic-circuit: circuitglobe.com, Retrieved 7 April, 2019

- Types-of-magnetic-materials: electronicspani.com, Retrieved 18 August, 2019

- What-is-self-inductance: circuitglobe.com, Retrieved 6 January, 2019

- Mutual-inductance, inductor: electronics-tutorials.ws, Retrieved 2 May, 2019

- What-is-a-magnetic-hysteresis: circuitglobe.com, Retrieved 13 August, 2019

5

Electrical Machines

The machines which convert electrical energy into mechanical energy or vice versa are termed as electrical machines. Some of the common electrical machines are electric generators, electric motors and transformers. The topics elaborated in this chapter will help in gaining a better perspective about these electrical machines.

An electrical machine is a device which converts mechanical energy into electrical energy or vice versa. Electrical machines also include transformers, which do not actually make conversion between mechanical and electrical form but they convert AC current from one voltage level to another voltage level.

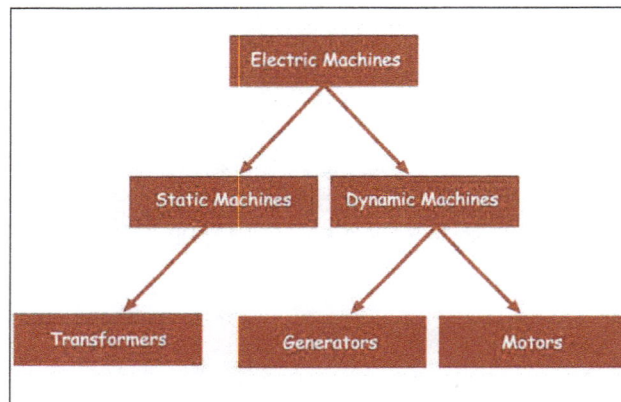

Electric Generator

Generator is a machine that converts mechanical energy into electrical energy. It works based on principle of faraday law of electromagnetic induction. The faradays law states that whenever a conductor is placed in a varying magnetic field, EMF is induced and this induced EMF is equal to the rate of change of flux linkages. This EMF can be generated when there is either relative space or relative time variation between the conductor and magnetic field. So the important elements of a generator are:

- Magnetic field
- Motion of conductor in magnetic field

Working of Generators

Generators are basically coils of electric conductors, normally copper wire, that are tightly wound onto a metal core and are mounted to turn around inside an exhibit of large magnets. An electric conductor moves through a magnetic field, the magnetism will interface with the electrons in the conductor to induce a flow of electrical current inside it.

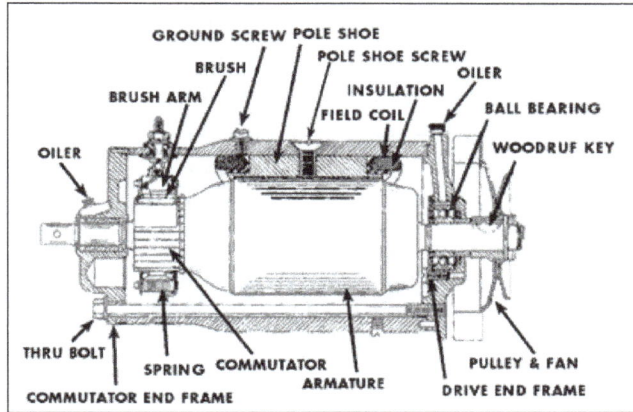

Typical Generator

The conductor coil and its core are called the armature, connecting the armature to the shaft of a mechanical power source, for example an motor, the copper conductor can turn at exceptionally increased speed over the magnetic field.

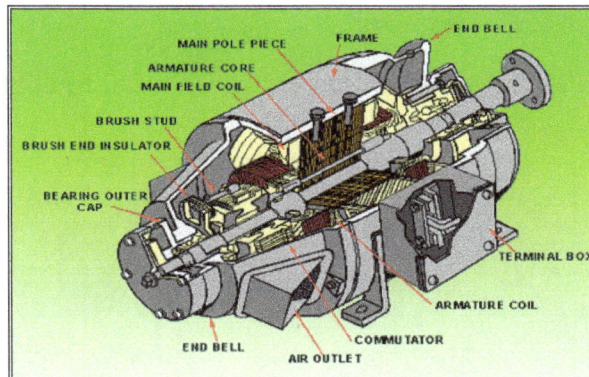

The point when the generator armature first starts to turn, then there is a weak magnetic field in the iron pole shoes. As the armature turns, it starts to raise voltage. Some of this voltage is making on the field windings through the generator regulator. This impressed voltage builds up stronger winding current, raises the strength of the magnetic field. The expanded field produces more voltage in the armature. This, in turn, make more current in the field windings, with a resultant higher armature voltage. At this time the signs of the shoes depended on the direction of flow of current in the field winding. The opposite signs will give current to flow in wrong direction.

Types of Generators

The generators are classified into types:

- AC generators

- DC generators

AC Generator

An alternating current (A.C.) generator is an important application of electromagnetic induction. A.C. generator is an electromagnetic device which transforms mechanical energy into electrical energy. It consists of a rectangular coil of wire which can be rotated about an axis. The coil is located between the poles of two permanent magnets. As the coil rotates, the magnetic field through the coil changes, which induces an electromotive force (e.m.f.) between the ends of the coil.

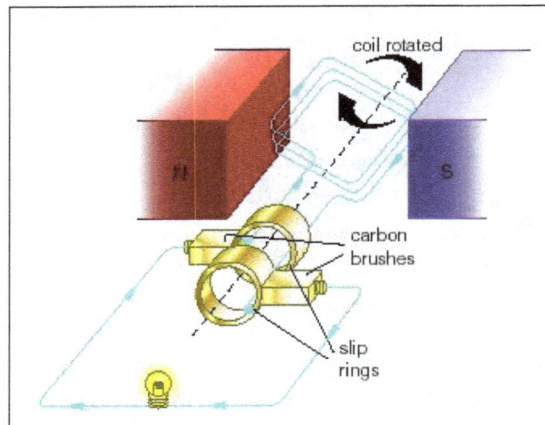

Working Principle of AC Generator

The working principle of an alternator or AC generator is similar to the basic working principle of a DC generator.

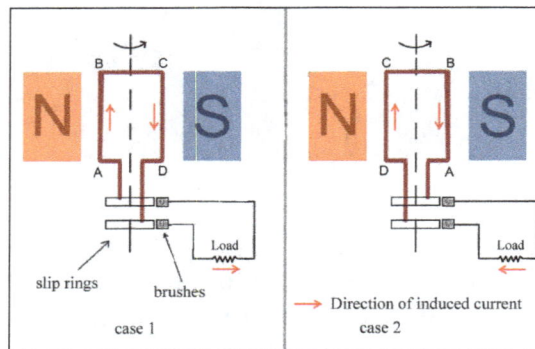

Figure helps you understanding how an alternator or AC generator works. According to the Faraday's law of electromagnetic induction, whenever a conductor moves in a magnetic field EMF gets induced across the conductor. If the close path is provided to the conductor, induced emf causes current to flow in the circuit.

Now, see the above figure. Let the conductor coil ABCD is placed in a magnetic field. The direction of magnetic flux will be form N pole to S pole. The coil is connected to slip rings, and the load is connected through brushes resting on the slip rings.

Now, consider the case 1 from above figure. The coil is rotating clockwise, in this case the direction of induced current can be given by Fleming's right hand rule, and it will be along A-B-C-D.

As the coil is rotating clockwise, after half of the time period, the position of the coil will be as in second case of above figure. In this case, the direction of the induced current according to Fleming's right hand rule will be along D-C-B-A. It shows that, the direction of the current changes after half of the time period that means we get an alternating current.

Construction of AC Generator (Alternator)

Main parts of the alternator, obviously, consists of stator and rotor. But, the unlike other machines, in most of the alternators, field exciters are rotating and the armature coil is stationary.

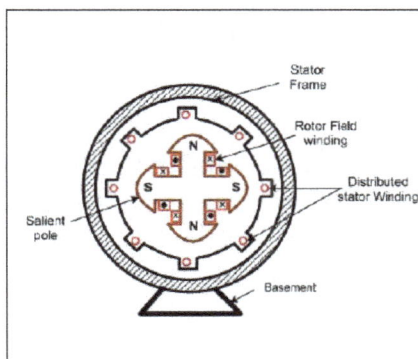

Salient pole type alternator.

Stator

Unlike in DC machine stator of an alternator is not meant to serve path for magnetic flux. Instead, the stator is used for holding armature winding. The stator core is made up of lamination of steel alloys or magnetic iron, to minimize the eddy current losses.

Why Armature Winding is Stationary in an Alternator?

- At high voltages, it easier to insulate stationary armature winding, which may be as high as 30 kV or more.

- The high voltage output can be directly taken out from the stationary armature. Whereas, for a rotary armature, there will be large brush contact drop at higher voltages, also the sparking at the brush surface will occur.

- Field exciter winding is placed in rotor, and the low dc voltage can be transferred safely.

- The armature winding can be braced well, so as to prevent deformation caused by the high centrifugal force.

Rotor

There are two types of rotor used in an AC generator / alternator:

- Salient

- Cylindrical type

- Salient pole type: Salient pole type rotor is used in low and medium speed alternators. Construction of AC generator of salient pole type rotor is shown in the figure. This type of rotor consists of large number of projected poles (called salient poles), bolted on a magnetic wheel. These poles are also laminated to minimize the eddy current losses. Alternators featuring this type of rotor are large in diameters and short in axial length.

- Cylindrical type: Cylindrical type rotors are used in high speed alternators, especially in turbo alternators. This type of rotor consists of a smooth and solid steel cylinder having slots along its outer periphery. Field windings are placed in these slots.

The DC supply is given to the rotor winding through the slip rings and and brushes arrangement.

Purpose of Slip Rings

The slip rings allow the transfer of alternating e.m.f. induced in the rotating coil to the external circuit. Each ring is connected to one end of the coil wire and is electrically connected to the external circuit via the conductive carbon brushes.

Note the difference between A.C. generator and D.C. motor. D.C. motor uses split-ring commutator, which reverses the current direction in the coil every half a turn and allows the coil to always turn in the clockwise direction.

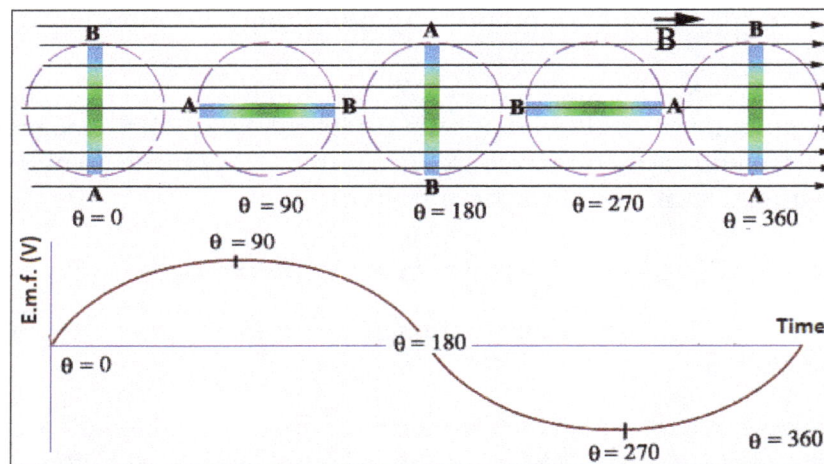

Using the figure, we will investigate the workings of an a.c. generator. Note that the coil is being turned in a clockwise manner and the magnetic field is pointing towards the right.

Steps in the Operation

- Coil starts in reference position 0°: The plane of the coil is perpendicular to the magnetic field lines. This means that the sides of the coil are moving parallel to the magnetic field lines and not "cutting" through any magnetic field lines. Hence, no e.m.f. is induced.

- Coil gets turned to reference position 90°: The plane of the coil is parallel to the magnetic field lines. The sides of the coil are moving perpendicularly to the magnetic field lines and will be "cutting" through the magnetic field lines at the greatest rate. Hence, the induced e.m.f. is the maximum at this position. Using Fleming's right hand rule, the direction of force at A is upwards (due to clockwise motion), while the magnetic field lines are pointing rightwards. This will give an induced current pointing into the screen. You can do the same analysis for B, which will be carrying an induced current pointing out of the screen.

- Coil gets turned to reference position 180° and 360°: Same as the analysis in reference position 0°.

- Coil gets turned to reference position 270°: Same analysis as in reference position 90° BUT the e.m.f. is in the opposite direction. This is due to the position of A and B switching places and by the Fleming's right hand rule, the inwards current will be carried by B and outwards current will be carried by A.

The frequency of rotation is related to the period T by:

$$f = \frac{1}{T}$$

Ways to increase emf in a.c. generator:

1. Decrease distance between magnet and coil. (To increase magnetic field strength experienced by coil).

2. Use a stronger magnet.

3. Increase frequency of rotation of the coil. (Double freq. = double max. e.m.f. and halving T).

4. Increase number of turns in the coil. (Double no. of turns = double max e.m.f.).

Turning the Magnets Instead of the Coil

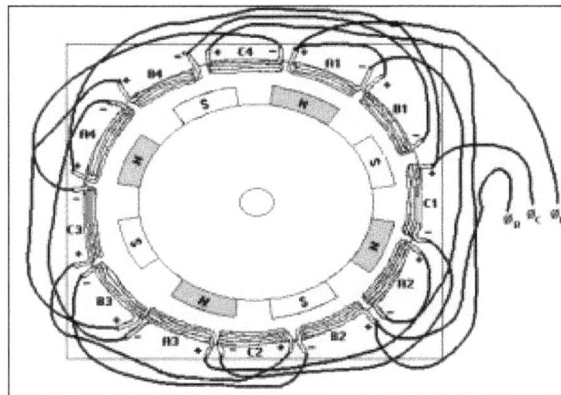

For the generation of large currents, it is more practical and advantageous to keep the coil fixed and to rotate the magnetic field around the coil. In this case, the magnetic field cuts the coil to produce the induced e.m.f., instead of the coil cutting the magnetic field. Note that the slip rings and carbon brushes (incapable of carrying large currents) are absent in this design.

DC Generator

A DC Generator is an electrical device which converts mechanical energy into electrical energy. It mainly consists of three main parts, i.e. Magnetic field system, Armature and Commutator and Brush gear. The other parts of a DC Generator are Magnetic frame and Yoke, Pole Core and Pole Shoes, Field or Exciting coils, Armature Core and Windings, Brushes, End housings, Bearings and Shafts.

The diagram of the main parts of a 4 pole DC Generator or DC Machine is shown below.

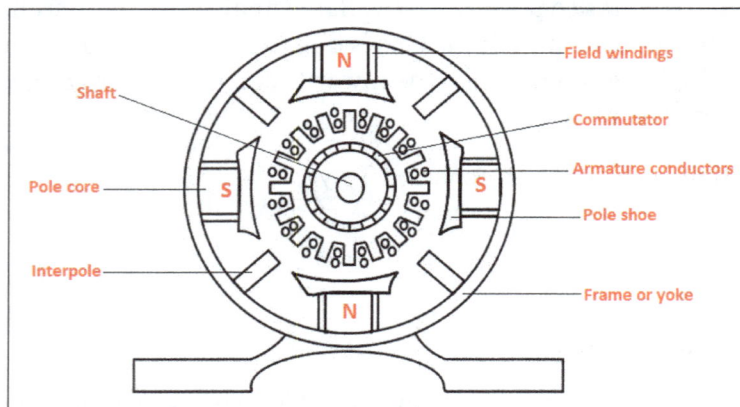

Magnetic Field System of DC Generator

The Magnetic Field System is the stationary or fixed part of the machine. It produces the main magnetic flux. The magnetic field system consists of Mainframe or Yoke, Pole core and Pole shoes and Field or Exciting coils.

Magnetic Frame and Yoke

The outer hollow cylindrical frame to which main poles and inter-poles are fixed and by means of which the machine is fixed to the foundation is known as Yoke. It is made of cast steel or rolled steel for the large machines and for the smaller size machine the yoke is generally made of cast iron.

The two main purposes of the yoke are as follows:

- It supports the pole cores and provides mechanical protection to the inner parts of the machines.

- It provides a low reluctance path for the magnetic flux.

Pole Core and Pole Shoes

The Pole Core and Pole Shoes are fixed to the magnetic frame or yoke by bolts. Since the poles, project inwards they are called salient poles. Each pole core has a curved surface. Usually, the pole core and shoes are made of thin cast steel or wrought iron laminations which are riveted together under hydraulic pressure. The poles are laminated to reduce the Eddy Current loss.

The figure of pole core and pole shoe are shown below.

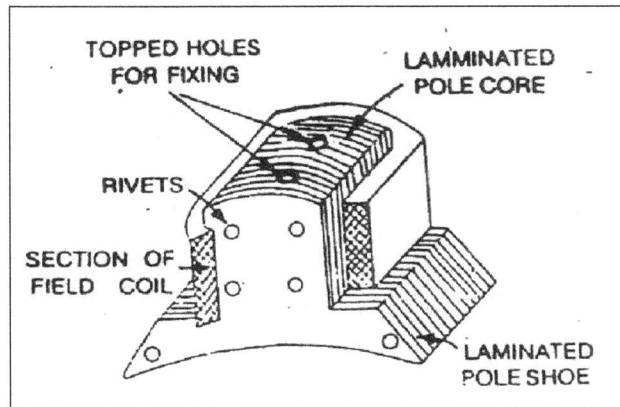

The poles core serves the following purposes given below:

- It supports the field or exciting coils.

- They spread out the magnetic flux over the armature periphery more uniformly.

- It increases the cross-sectional area of the magnetic circuit, as a result, the reluctance of the magnetic path is reduced.

Field or Exciting Coils

Each pole core has one or more field coils (windings) placed over it to produce a magnetic field. The enamelled copper wire is used for the construction of field or exciting coils. The coils are wound on the former and then placed around the pole core.

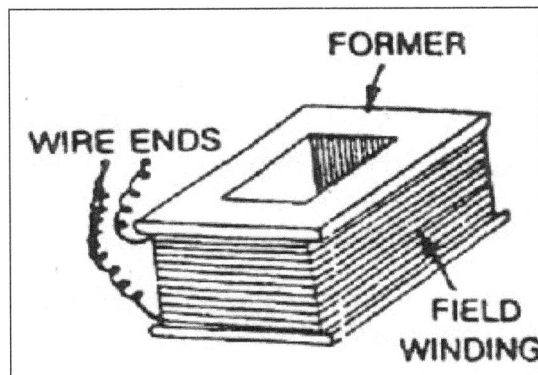

When direct current passes through the field winding, it magnetizes the poles, which in turns produces the flux. The field coils of all the poles are connected in series in such a way that when current flows through them, the adjacent poles attain opposite polarity.

Armature of DC Generator

The rotating part of the DC machine or a DC Generator is called the Armature. The armature consists of a shaft upon which a laminated cylinder, called Amature Core is placed.

Armature Core

The armature core of DC Generator is cylindrical in shape and keyed to the rotating shaft. At the outer periphery of the armature have grooves or slots which accommodate the armature winding as shown in the figure.

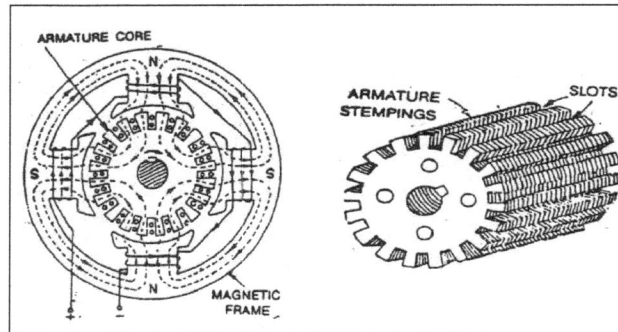

The armature core of a DC generator or machine serves the following purposes:

- It houses the conductors in the slots.

- It provides an easy path for the magnetic flux.

As the armature is a rotating part of the DC Generator or machine, the reversal of flux takes place in the core, hence hysteresis losses are produced. The silicon steel material is used for the construction of the core to reduce the hysteresis losses.

The rotating armature cuts the magnetic field, due to which an emf is induced in it. This emf circulates the eddy current which results in Eddy Current loss. Thus to reduce the loss the armature core is laminated with a stamping of about 0.3 to 0.5 mm thickness. Each lamination is insulated from the other by a coating of varnish.

Armature Winding

The insulated conductors are placed in the slots of the armature core. The conductors are wedged, and bands of steel wire wound around the core and are suitably connected. This arrangement of conductors is called Armature Winding. The armature winding is the heart of the DC Machine.

Armature winding is a place where conversion of power takes place. In the case of a DC Generator here, mechanical power is converted into electrical power. On the basis of connections, the windings are classified into two types named as Lap Winding and Wave Winding.

Lap Winding

In lap winding, the conductors are connected in such a way that the numbers of parallel paths are equal to the number of poles. Thus, if a machine has P poles and Z armature conductors, then there will be P parallel paths; each path will have Z/P conductors connected in series.

In lap winding, the number of brushes is equal to the number of parallel paths. Out of which half the brushes are positive and the remaining halves are negative.

Wave Winding

In wave winding, the conductors are so connected that they are divided into two parallel paths irrespective of the number of poles of the machine. Thus, if the machine has Z armature conductors, there will be only two parallel paths each having Z/2 conductors in series. In this case number of brushes is equal to two, i.e. number of parallel paths.

Commutator in DC Generator

The commutator, which rotates with the armature, is cylindrical in shape and is made from a number of wedge-shaped hard drawn copper bars or segments insulated from each other and from the shaft. The segments form a ring around the shaft of the armature. Each commutator segment is connected to the ends of the armature coils.

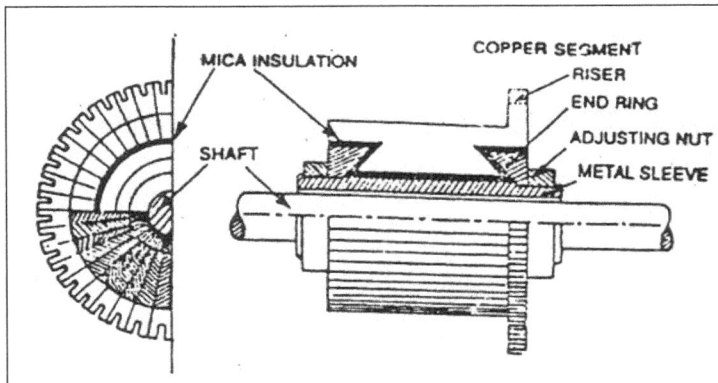

It is the most important part of a DC machine and serves the following purposes;

- It connects the rotating armature conductors to the stationary external circuit through brushes.

- It converts the induced alternating current in the armature conductor into unidirectional current in the external load circuit in DC Generator action, whereas it converts the alternating torque into unidirectional (continuous) torque produced in the armature in motor action.

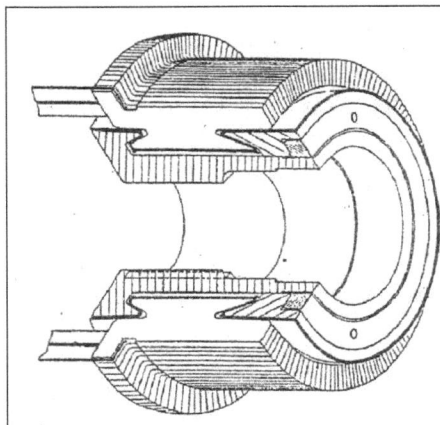

Brushes

Carbon brushes are placed or mounted on the commutator and with the help of two or more carbon brushes current is collected from the armature winding. Each brush is supported in a metal box called a brush box or brush holder. The brushes are pressed upon the commutator and form the connecting link between the armature winding and the external circuit.

The pressure exerted by the brushes on the commutator can be adjusted and is maintained at a constant value by means of springs. With the help of the brushes the current which is produced on the windings, is passed on to the commutator and then to the external circuit.

They are usually made of high-grade carbon because carbon is conducting material and at the same time in powdered form provides a lubricating effect on the commutator surface.

End Housings

End housings are attached to the ends of the Mainframe and provide support to the bearings. The front housings support the bearing and the brush assemblies whereas the rear housings usually support the bearings only.

Bearings

The ball or roller bearings are fitted in the end housings. The function of the bearings is to reduce friction between the rotating and stationary parts of the machine. Mostly high carbon steel is used for the construction of bearings as it is very hard material.

Shaft

The shaft is made of mild steel with a maximum breaking strength. The shaft is used to transfer mechanical power from or to the machine. The rotating parts like armature core, commutator, cooling fans, etc. are keyed to the shaft.

Working Principle of a DC Generator

According to Faraday's laws of electromagnetic induction, whenever a conductor is placed in a varying magnetic field (OR a conductor is moved in a magnetic field), an emf (electromotive force) gets induced in the conductor. The magnitude of induced emf can be calculated from the emf equation of dc generator. If the conductor is provided with a closed path, the induced current will circulate within the path. In a DC generator, field coils produce an electromagnetic field and the armature conductors are rotated into the field. Thus, an electromagnetically induced emf is generated in the armature conductors. The direction of induced current is given by Fleming's right hand rule.

Need of a Split Ring Commutator

According to Fleming's right hand rule, the direction of induced current changes whenever the direction of motion of the conductor changes. Let's consider an armature rotating clockwise and a conductor at the left is moving upward. When the armature completes a half rotation, the direction of motion of that particular conductor will be reversed to downward. Hence, the direction of current in every armature conductor will be alternating. If you look at the above figure, you will know how the direction of the induced current is alternating in an armature conductor. But with a split ring commutator, a connection of the armature conductors also gets reversed when the current reversal occurs. And therefore, we get unidirectional current at the terminals.

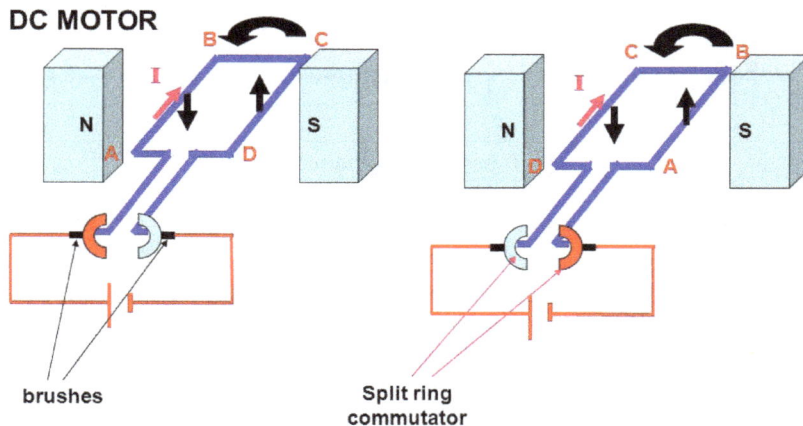

DC MOTOR

brushes

Split ring commutator

DC Generator E.M.F Equation

The emf equation of dc generator according to Faraday's Laws of Electromagnetic Induction is:

$$E_g = P\emptyset ZN / 60 \ A$$

Where,

- $\emptyset$ is a flux or pole within Webber.

- Z is a total no.of armature conductor.

- P is a number of poles in a generator.

- A is a number of parallel lanes within the armature.

- N is the rotation of armature in r.p.m (revolutions per minute).

- E is the induced e.m.f in any parallel lane within the armature.

- E_g is the generated e.m.f in any one of the parallel lanes.

N/60 is the number of turns per second.

Time for one turn will be dt = 60/N sec.

Types of DC Generator

The classification of DC generators can be done in two most important categories namely separately excited as well as self-excited.

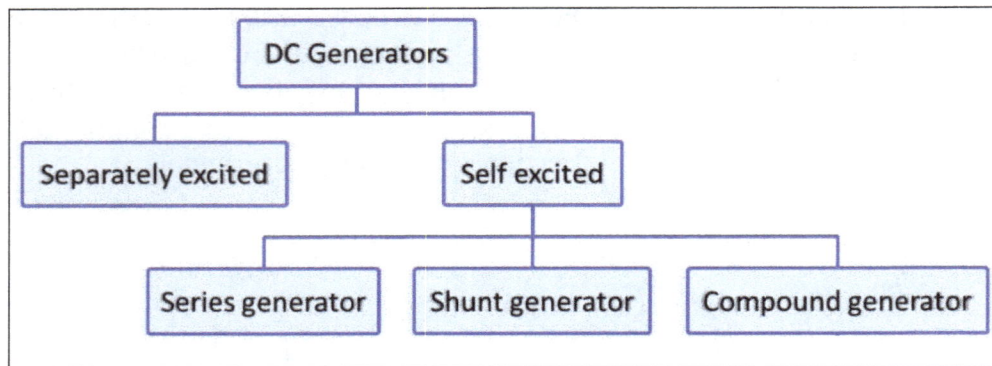

Types of DC Generators

Separately Excited

In separately excited type, the field coils are strengthened from an autonomous exterior DC source.

Self Excited

In self-excited type, the field coils are strengthened from the generated current with the generator. The generation of first electromotive force will occur because of its outstanding magnetism within field poles.

The produced electromotive force will cause a fraction of current to supply in the field coils, therefore which will increase the field flux as well as electromotive force generation. Further, these types of dc generators can be classified into three types namely series wound, shunt wound, and compound wound.

- In a series wound, both the field winding & armature winding are connected in series with each other.

- In shunt wound, both the field winding & armature winding are connected in parallel with each other.

- The compound winding is the blend of series winding & shunt winding.

Applications of DC Generators

The applications of different types of DC generator include the following.

- The separately excited type DC generator is used for boosting as well as electroplating. It is used in power and lighting purpose using field regulator

- The self-excited DC generator or shunt DC generator is used for power as well as ordinary lighting using the regulator. It can be used for battery lighting.

- The series DC generator is used in arc lamps for lighting, stable current generator and booster.

- Compound DC generator is used to provide the power supply for DC welding machines.

- Level compound DC generator is used to provide a power supply for hostels, lodges, offices, etc.

- Over compound, DC generator is used to reimburse the voltage drop within Feeders.

Electric Motor

The motor or an electrical motor is a device that has brought about one of the biggest advancements in the fields of engineering and technology ever since the invention of electricity. A motor is nothing but an electro-mechanical device that converts electrical energy into mechanical energy. It's because of motors, life is what it is today in the 21st century. Without the motor, we had still been living in Sir Thomas Edison's Era where the only purpose of electricity would have been to glow bulbs. There are different types of motor have been developed for different specific purposes.

In simple words, we can say a device that produces rotational force is a motor. The very basic principle of functioning of an electrical motor lies on the fact that force is experienced in the direction perpendicular to the magnetic field and the current, when field and current are made to interact with each other.

Ever since the invention of motors, a lot of advancements has taken place in this field of engineering and it has become a subject of extreme importance for modern engineers.

Classification or Types of Motor

The primary classification of motor or types of motor can be tabulated as shown below,

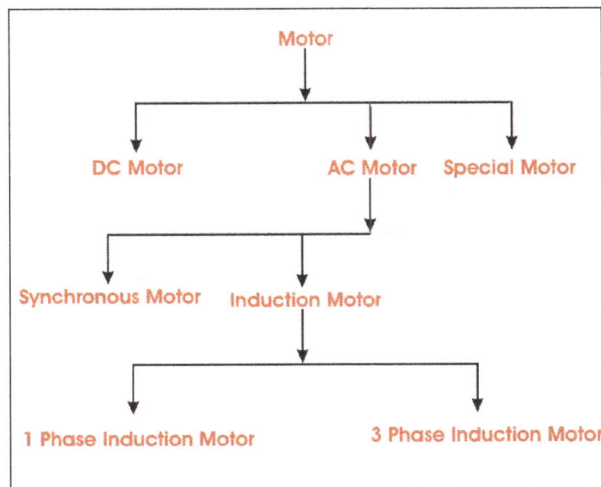

AC Motor

The motor that converts the alternating current into mechanical power by using an electromagnetic induction phenomenon is called an AC motor. This motor is driven by an alternating current. The stator and the rotor are the two most important parts of the AC motors. The stator is the stationary part of the motor, and the rotor is the rotating part of the motor. The AC motor may be single phase or three phases.

The three phase AC motors are mostly applied in the industry for bulk power conversion from electrical to mechanical. For small power conversion, the single phase AC motors are mostly used. The single phase AC motor is nearly small in size, and it provides a variety of services in the home, office, business concerns, factories, etc. Almost all the domestic appliances such as refrigerators, fans, washing machine, hair dryers, mixers, etc., use single phase AC motor.

The AC motor is mainly classified into two types. They are the synchronous motor and the induction motor.

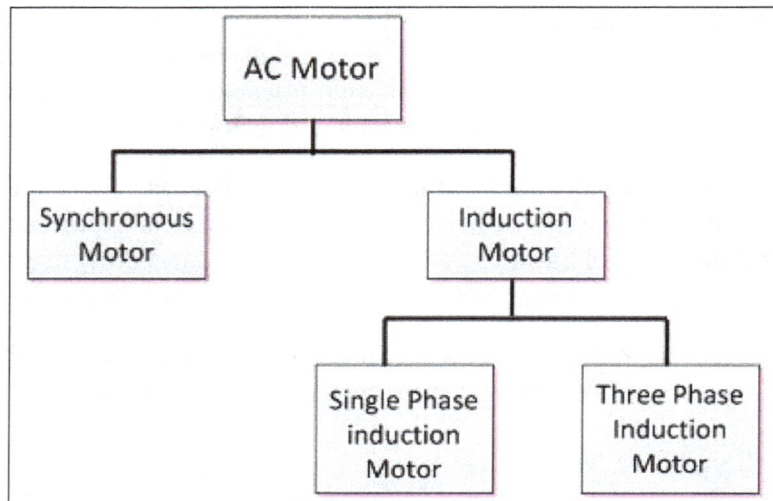

Synchronous Motor

The motor that converts the AC electrical power into mechanical power and is operated only at the synchronous speed is known as a synchronous motor.

Working Principle of a Synchronous Motor

When supply is given to synchronous motor, a revolving field is set up. This field tries to drag the rotor with it, but could not do so because of rotor inertia. Hence, no starting torque is produced. Thus, inherently synchronous motor is not a self-starting the motor.

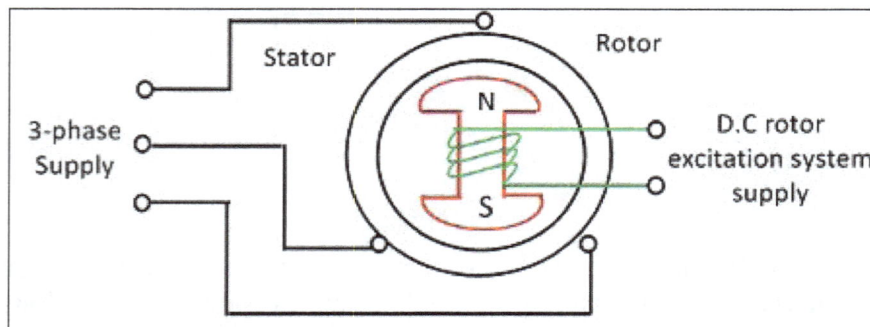

Induction Motor

An induction motor (also known as an asynchronous motor) is a commonly used AC electric motor. In an induction motor, the electric current in the rotor needed to produce torque is obtained via electromagnetic induction from the rotating magnetic field of the stator winding. The rotor of an induction motor can be a squirrel cage rotor or wound type rotor.

Induction motors are referred to as 'asynchronous motors' because they operate at a speed less than their synchronous speed. So first thing to understand – what is synchronous speed?

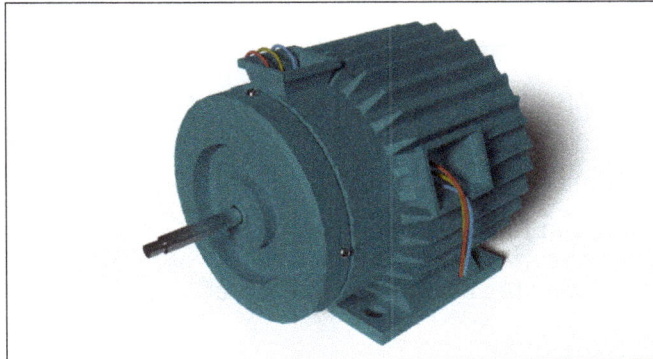

A Typical Induction Motor.

Synchronous Speed

Synchronous speed is the speed of rotation of the magnetic field in a rotary machine, and it depends upon the frequency and number poles of the machine. The induction motor always runs at speed less than its synchronous speed. The rotating magnetic field produced in the stator will create flux in the rotor, hence causing the rotor to rotate. Due to the lag between the flux current in the rotor and the flux current in the stator, the rotor will never reach its rotating magnetic field speed (i.e. the synchronous speed).

There are basically two types of induction motor. The types of induction motor depend upon the input supply. There are single phase induction motors and three phase induction motors. Single phase induction motors are not a self-starting motor, and three phase induction motor are a self-starting motor.

Working Principle of Induction Motor

We need to give double excitation to make a DC motor to rotate. In the DC motor, we give one supply to the stator and another to the rotor through brush arrangement. But in induction motor, we give only one supply, so it is interesting to know how an induction motor works. It is simple, from the name itself we can understand that here, the induction process is involved. When we give the supply to the stator winding, a magnetic flux gets produced in the stator due to the flow of current in the coil. The rotor winding is so arranged that each coil becomes short-circuited.

The flux from the stator cuts the short-circuited coil in the rotor. As the rotor coils are short-circuited, according to Faraday's law of electromagnetic induction, the current will start flowing through the coil of the rotor. When the current through the rotor coils flows, another flux gets generated in the rotor. Now there are two fluxes, one is stator flux, and another is rotor flux. The rotor flux will be lagging in respect of the stator flux. Because of that, the rotor will feel a torque which will make the rotor to rotate in the direction of the rotating magnetic field. This is the working principle of both single and three phase induction motors.

Types of Induction Motors

The types of induction motors can be classified depending on whether they are a single phase or three phase induction motor.

Single Phase Induction Motor

The types of single phase induction motors include:

1. Split Phase Induction Motor

2. Capacitor Start Induction Motor

3. Capacitor Start and Capacitor Run Induction Motor

4. Shaded Pole Induction Motor

Three Phase Induction Motor

The types of three phase induction motors include:

1. Squirrel Cage Induction Motor

2. Slip Ring Induction Motor

The single-phase induction motor is not a self-starting motor, and that the three-phase induction motor is self-starting.

When the motor starts running automatically without any external force applied to the machine, then the motor is referred to as 'self-starting'. For example, we see that when we put on the switch the fan starts to rotate automatically, so it is a self-starting machine. Point to be noted that fan used in home appliances is a single phase induction motor which is inherently not self-starting.

Three Phase Induction Motor

In a three phase system, there are three single phase lines with a 120° phase difference. So the rotating magnetic field has the same phase difference which will make the rotor to move. If we consider three phases a, b, and c when phase a gets magnetized, the rotor will move towards the phase a winding a, in the next moment phase b will get magnetized and it will attract the rotor and then phase c. So the rotor will continue to rotate.

Single Phase Induction Motor

It has only one phase still it makes the rotor to rotate, so it is quite interesting. Before that, we need to know why a single phase induction motor is not a self-starting motor and how we overcome the problem. We know that the AC supply is a sinusoidal wave and it produces a pulsating magnetic field in the uniformly distributed stator winding.

Since we can assume the pulsating magnetic field as two oppositely rotating magnetic fields, there will be no resultant torque produced at the starting, and hence the motor does not run. After giving the supply, if the rotor is made to rotate in either direction by an external force, then the motor will start to run. We can solve this problem by making the stator winding into two winding – one is the main winding, and another is auxiliary winding.

We connect one capacitor in series with the auxiliary winding. The capacitor will make a phase difference when current flows through both coils. When there is a phase difference, the rotor will generate a starting torque, and it will start to rotate. Practically we can see that the fan does not rotate when the capacitor gets disconnected from the motor, but if we rotate with the hand, it will start rotating. That is why we use a capacitor in the single-phase induction motor.

Due to the various advantages of an induction motor, there is a wide range of applications of an induction motor. One of their biggest advantages is their high efficiency – which can go as high as 97%. The main disadvantage of an induction motor is that the speed of the motor varies with the applied load. The direction of rotation of induction motor can easily be changed by changing the phase sequence of three-phase supply, i.e., if RYB is in a forward direction, the RBY will make the motor to rotate in reverse direction. This is in the case of three phase motor, but in a single phase motor, the direction can be reversed by reversing the capacitor terminals in the winding.

DC Motor

A DC motor is an electric motor that runs on direct current power. In any electric motor, operation is dependent upon simple electromagnetism. A current carrying conductor generates a magnetic field, when this is then placed in an external magnetic field; it will encounter a force proportional to the current in the conductor and to the strength of the external magnetic field. It is a device which converts electrical energy to mechanical energy. It works on the fact that a current carrying conductor placed in a magnetic field experiences a force which causes it to rotate with respect to its original position.

Practical DC Motor consists of field windings to provide the magnetic flux and armature which acts as the conductor.

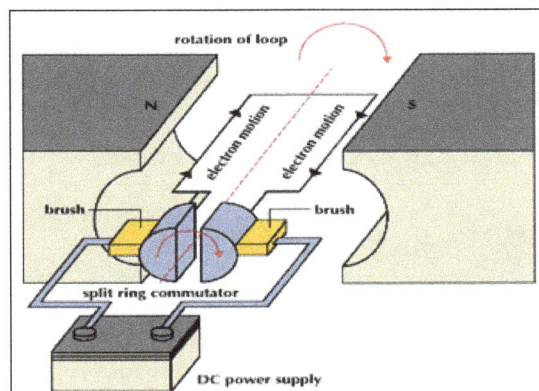

Brushless DC Motors Work.

The input of a brushless DC motor is current/voltage and its output is torque. Understanding the operation of DC motor is very simple from a basic diagram is shown in below. DC motor basically consist two main parts. The rotating part is called the rotor and the stationary part is also called the stator. The rotor rotates with respect to the stator.

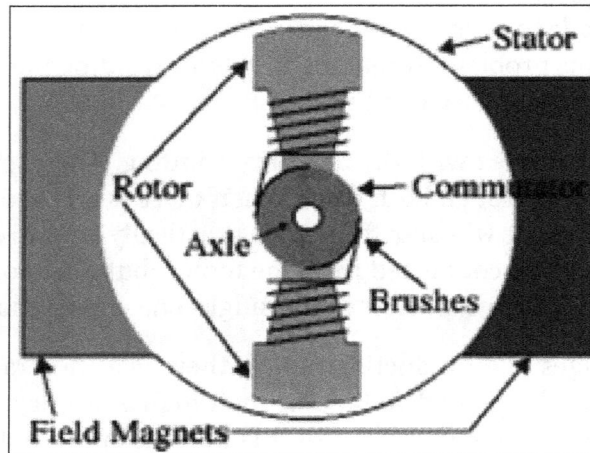

Dc Motor.

The rotor consists of windings, the windings being electrically associated with the commutator. The geometry of the brushes, commutator contacts and rotor windings are such that when power is applied, the polarities of the energized winding and the stator magnets are misaligned and the rotor will turn until it is very nearly straightened with the stator's field magnets.

As the rotor reaches alignment, the brushes move to the next commutator contacts and energize the next winding. The rotation reverses the direction of current through the rotor winding, prompting a flip of the rotor's magnetic field, driving it to keep rotating.

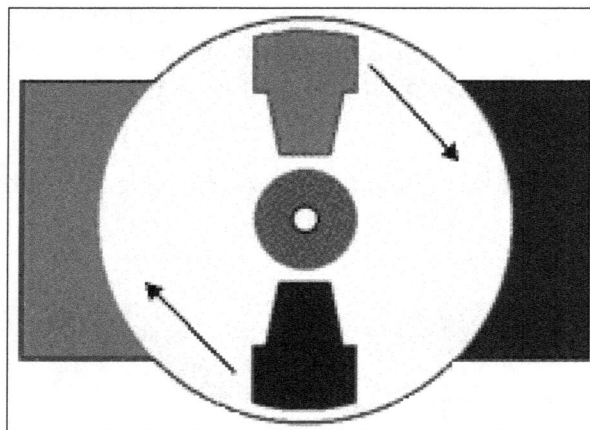

Advantages of DC Motor

1. Provide excellent speed control for acceleration and deceleration.

2. Easy to understand design.

3. Simple, cheap drive design.

Principle of DC Motor

When a current carrying conductor is placed in a magnetic field, it experiences a torque and has a tendency to move. In other words, when a magnetic field and an electric field interact, a mechanical force is produced. The DC motor or direct current motor works on that principal. This is known as motoring action.

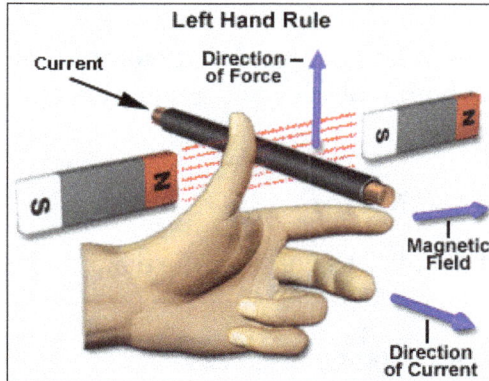

The direction of rotation of a this motor is given by Fleming's left hand rule, which states that if the index finger, middle finger, and thumb of your left hand are extended mutually perpendicular to each other and if the index finger represents the direction of magnetic field, middle finger indicates the direction of current, then the thumb represents the direction in which force is experienced by the shaft of the DC motor.

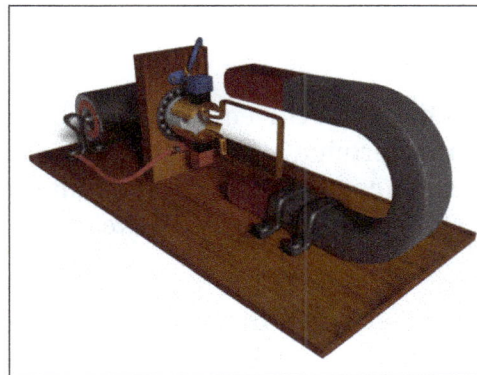

Structurally and construction wise a direct current motor is exactly similar to a DC generator, but electrically it is just the opposite. Here we unlike a generator we supply electrical energy to the input port and derive mechanical energy from the output port. We can represent it by the block diagram shown below.

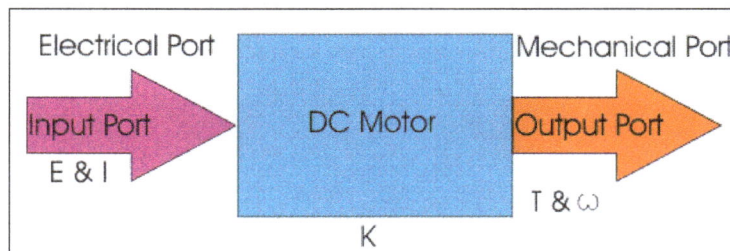

Here in a DC motor, the supply voltage E and current I is given to the electrical port or the input port and we derive the mechanical output i.e. torque T and speed ω from the mechanical port or output port.

The parameter K relates the input and output port variables of the direct current motor.

$T = KI$ and $E = K\omega$

So from the picture above, we can well understand that motor is just the opposite phenomena of a DC generator, and we can derive both motoring and generating operation from the same machine by simply reversing the ports.

Detailed Description of a DC Motor

To understand the DC motor in details let's consider the diagram below,

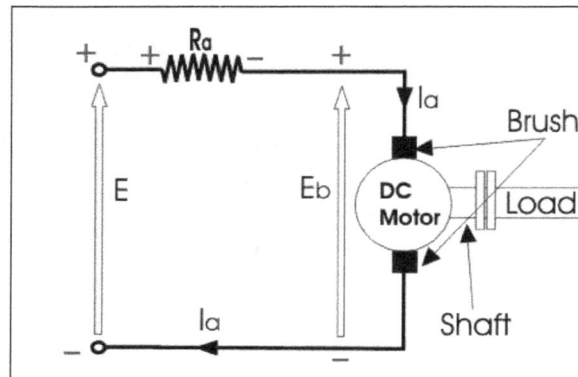

The circle in the center represents the direct current motor. On the circle, we draw the brushes. On the brushes, we connect the external terminals, through which we give the supply voltage. On the mechanical terminal, we have a shaft coming out from the center of the armature, and the shaft couples to the mechanical load. On the supply terminals, we represent the armature resistance R_a in series.

Now, let the input voltage E, is applied across the brushes. Electric current which flows through the rotor armature via brushes, in presence of the magnetic field, produces a torque T_g. Due to this torque T_g the dc motor armature rotates. As the armature conductors are carrying currents and the armature rotates inside the stator magnetic field, it also produces an emf E_b in the manner very similar to that of a generator. The generated Emf E_b is directed opposite to the supplied voltage and is known as the back Emf, as it counters the forward voltage.

The back emf like in case of a generator is represented by,

$$E_b = \frac{P.\varphi.Z.N}{60.A}$$

Where,

- P = no of poles,

- φ = flux per pole,
- Z= No. of conductors,
- A = No. of parallel paths,
- N is the speed of the DC Motor.

So, from the above equation, we can see E_b is proportional to speed 'N.' That is whenever a direct current motor rotates; it results in the generation of back Emf. Now let's represent the rotor speed by ω in rad/sec. So E_b is proportional to ω.

So, when the application of load reduces the speed of the motor, E_b decreases. Thus the voltage difference between supply voltage and back emf increases that means $E - E_b$ increases. Due to this increased voltage difference, the armature current will increase and therefore torque and hence speed increases. Thus a DC Motor is capable of maintaining the same speed under variable load.

Now armature current I_a is represented by,

$$I_a = \frac{E - E_b}{R_a}$$

Now at starting, speed $\omega = 0$ so at starting $E_b = 0$.

$$\therefore I_a = \frac{E}{R_a}$$

Now since the armature winding electrical resistance R_a is small, this motor has a very high starting current in the absence of back Emf. As a result we need to use a starter for starting a DC Motor.

Now as the motor continues to rotate, the back emf starts being generated and gradually the current decreases as the motor picks up speed.

4 Quadrant Operation of DC Motor

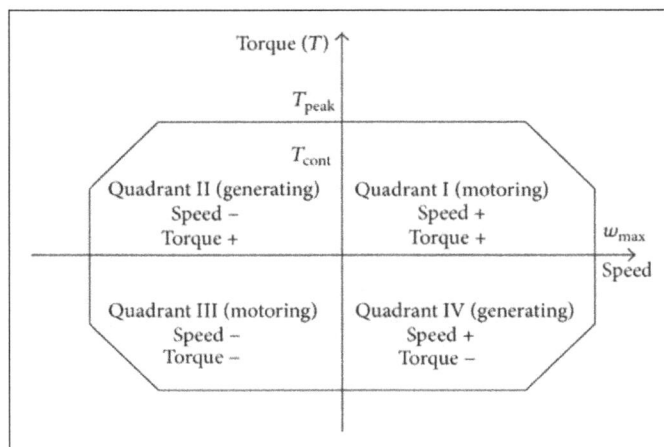

4 Quadrant Operation of DC Motor

Generally a motor can operate in 4 different regions:

- As a motor in forward or clockwise direction.

- As a generator in forward direction.

- As a motor in reverse or anticlockwise direction.

- As a generator in reverse direction.

- In the first quadrant, motor is driving the load with both the speed and torque in positive direction.

- In the second quadrant, torque direction reverses and motor acts as a generator.

- In the third quadrant, motor drives the load with speed and torque in negative direction.

- In the 4th quadrant, motor acts a generator in reverse mode.

In the first and third quadrant, motor acts in both forward and reverse directions. For example, motors in cranes to lift the load and also put it down.

In the second and fourth quadrant, motor acts as a generator in forward and reverse directions respectively and provides energy back to the power source. Thus the way to control a motor operation, to make it operate in any of the 4 quadrants is by controlling its speed and direction of rotation. Speed is controlled either by varying the armature voltage or weakening the field. The torque direction or direction of rotation is controlled by varying the extent to which applied voltage is greater than or less than the back emf.

Types of DC Motors

Geared DC Motors

Geared motors tend to reduce the speed of the motor but with a corresponding increase in torque. This property comes in handy, as DC motors can rotate at speeds much too fast for an electronic device to makes use of. Geared motors commonly consist of a DC brush motor and a gearbox attached to the shaft. Motors are distinguished as a geared by two connected units. It has many applications due to its cost of designing, reduces the complexity and constructing applications such as industrial equipment, actuators, medical tools and robotics.

- No good robot can ever be built without gears. All things considered, a good understanding of how gears affect parameters such as torque and velocity are very important.

- Gears work on the principle of mechanical advantage. This implies that by using distinctive gear diameters, we can exchange between rotational velocity and torque. Robots do not have a desirable speed to torque ratio.

- In robotics, torque is better than speed. With gears, it is possible to exchange the high velocity with a better torque. The increase in torque is inversely proportional to the reduction in speed.

Geared DC Motors.

Speed Reduction in Geared DC Motor

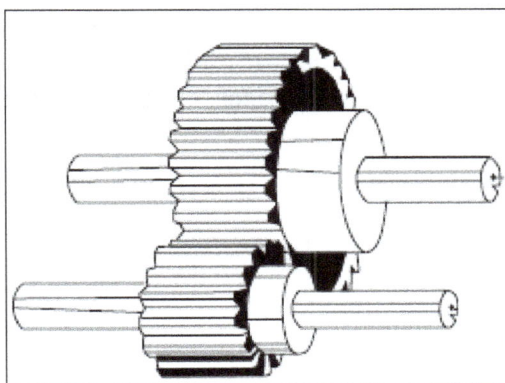

Speed Reduction in geared DC Motor.

Speed reduction in gears comprises of a little gear driving a larger gear. There may be few sets of these reduction gear sets in a reduction gear box. Sometimes the objective of using a gear motor is to reduce the rotating shaft speed of a motor in the device being driven, for example in a small electric clock where the tiny synchronous motor may be turning at 1,200 rpm however is decreased to one rpm to drive the second hand and further reduced in the clock mechanism to drive the minute and hour hands. Here the amount of driving force is irrelevant as long as it is sufficient to overcome the frictional impacts of the clock mechanism.

Series DC Motor

Series motor is a DC series motor where field winding is connected internally in series to the armature winding. The series motor provides high starting torque but must never be run without a load and is able to move very large shaft loads when it is first energized. Series motors are also known as series-wound motor.

In series motors, the field windings are associated in series with the armature. The field strength varies with progressions in armature current. At the time its speed is reduced by a load, the series motor advances more excellent torque. Its starting torque is more than different sorts of DC motor. It can also radiate more easily the heat that has built up in the winding due to the large amount of current being carried. Its speed shifts considerably between full-load and no-load. When load is removed, motor speed increases and current through the armature and field coils decreases. Unloaded operation of large machines is hazardous.

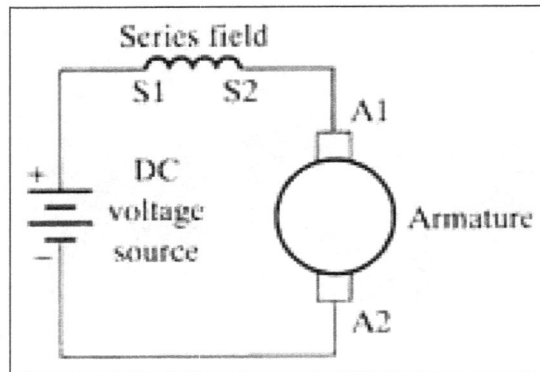

Series Motor.

Current through the armature and field coils decreases, the strength of the flux lines around them weakens. If the strength of the flux lines around the coils were reduced at the same rate as the current flowing through them, both would decrease at the same rate which the motor speed increases.

Advantages of Series Motor

- Huge starting torque

- Simple Construction

- Designing is easy

- Maintenance is easy

- Cost effective

Applications of Series Motor

Series Motors can produce enormous turning power, the torque from its idle state. This characteristic makes series motors suitable for small electrical appliances, versatile electric equipments and etc. Series motors are not suitable when a constant speed is needed. The reason is that the velocity of series motors varies greatly with varying load.

Shunt Motor

Shunt motors are shunt DC motors, where the field windings shunted to or are connected in parallel to the armature winding of the motor. The shunt DC motor is commonly used because of its best speed regulation. Also hence both the armature winding and the field windings are presented to the same supply voltage, however there are discrete branches for the stream of armature current and the field current.

A shunt motor has somewhat distinctive working characteristics than a series motor. Since the shunt field coil is made of fine wire, it cannot produce the large current for starting like the series field. This implies that the shunt motor has extremely low starting torque, which requires that the shaft load be quite little.

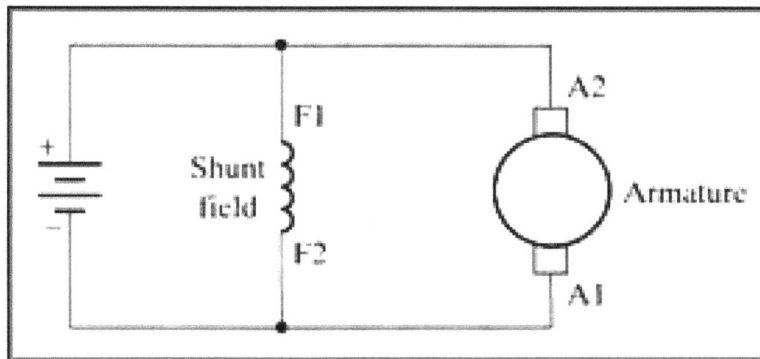

Shunt Motor.

When voltage is applied to the shunt motor, a very low amount of current flow through the shunt coil. The armature for the shunt motor is similar to the series motor and it will draw current to produce a strong magnetic field. Due to the interaction of magnetic field around armature and the field produced around the shunt field, the motor starts to rotate. Like the series motor, when the armature begins to turn, it will produce back EMF. The back EMF will cause the current in the armature to begin to diminish to a very small level. The amount of current the armature will draw is directly related to the size of the load when the motor reaches full speed. Since the load is generally small, the armature current will be small.

Advantages of Shunt Motor

- Simple control performance, resulting in a high level of flexibility for solving complex drive problems.

- High availability, therefore minimal service effort needed.

- High level of electro-magnetic compatibility.

- Very smooth running, therefore low mechanical stress of the overall system and high dynamic control processes.

- Wide control range and low speeds, therefore universally usable.

Applications of Shunt Motor

Shunt DC motors are very suitable for belt-driven applications. This constant speed motor is used in industrial and automotive applications such as machine tools and winding/unwinding machines where great amount of torque precision is required.

Transformer

A Transformer is a static electrical machine which transfers AC electrical power from one circuit to the other circuit at the constant frequency, but the voltage level can be altered that means voltage can be increased or decreased according to the requirement.

It works on the principle of Faraday's Law of Electromagnetic Induction which states that "the magnitude of voltage is directly proportional to the rate of change of flux."

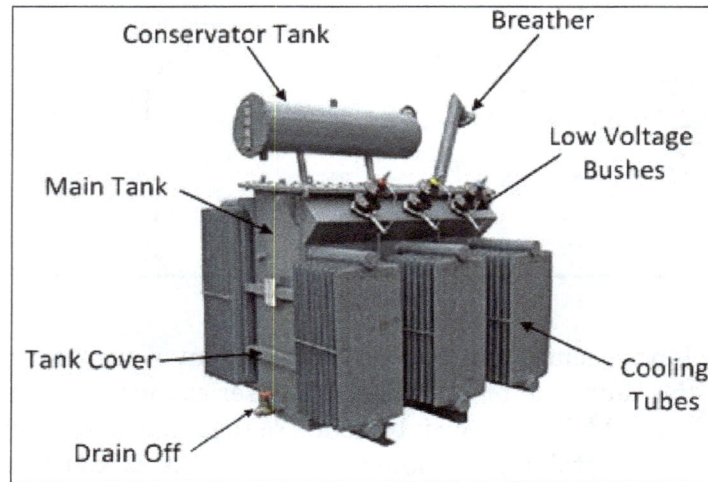

Necessity of a Transformer

Usually, electrical power is generated at 11Kv. For economical reasons AC power is transmitted at very high voltages say 220 kV or 440 kV over long distances. Therefore a step-up transformer is applied at the generating stations. Now for safety reasons the voltage is stepped down to different levels by step down transformer at various substations to feed the power to the different locations and thus the utilisation of power is done at 400/230 V.

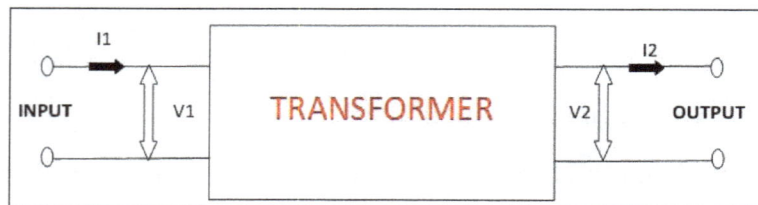

1. If $(V_2 > V_1)$ the voltage is raised on the output side and is known as Step-up transformer.

2. If $(V_2 < V_1)$ the voltage level is lowered on the output side and is known as Step down transformer.

Construction of a Transformer

It mainly consists of:

1. Magnetic circuit (consisting of core, limbs, yoke and damping structure.

2. Electrical circuit (consists of primary and secondary windings).

3. Dielectric circuit (consisting of insulations in different forms and used at the different places).

4. Tanks and accessories (conservator, breather, bushings, cooling tubes, etc.).

Working Principle of Transformer

The working principle of a transformer is very simple. Mutual induction between two or more windings (also known as coils) allows for electrical energy to be transferred between circuits.

Transformer Theory

Say you have one winding (also known as a coil) which is supplied by an alternating electrical source. The alternating current through the winding produces a continually changing and alternating flux that surrounds the winding. If another winding is brought close to this winding, some portion of this alternating flux will link with the second winding. As this flux is continually changing in its amplitude and direction, there must be a changing flux linkage in the second winding or coil.

According to Faraday's law of electromagnetic induction, there will be an EMF induced in the second winding. If the circuit of this secondary winding is closed, then a current will flow through it. This is the basic working principle of a transformer. Let us use electrical symbols to help visualize this. The winding which receives electrical power from the source is known as the 'primary winding'. In the diagram below this is the 'First Coil'.

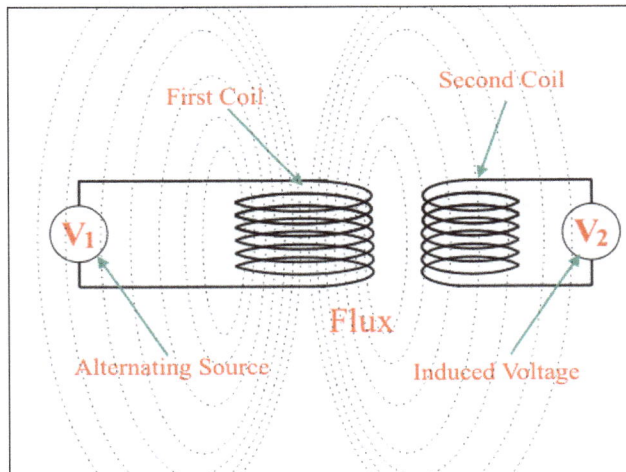

The winding which gives the desired output voltage due to mutual induction is commonly known as the 'secondary winding'. This is the 'Second Coil' in the diagram.

A transformer that increases voltage between the primary to secondary windings is defined as a step-up transformer. Conversely, a transformer that decreases voltage between the primary to secondary windings is defined as a step-down transformer.

While the diagram of the transformer above is theoretically possible in an ideal transformer – it is not very practical. This is because in open air only a very tiny portion of the flux produced from the first coil will link with the second coil. So the current that flows through the closed circuit connected to the secondary winding will be extremely small (and difficult to measure).

The rate of change of flux linkage depends upon the amount of linked flux with the second winding. So ideally almost all of the flux of primary winding should link to the secondary winding. This is effectively and efficiently done by using a core type transformer. This provides a low reluctance path common to both of the windings.

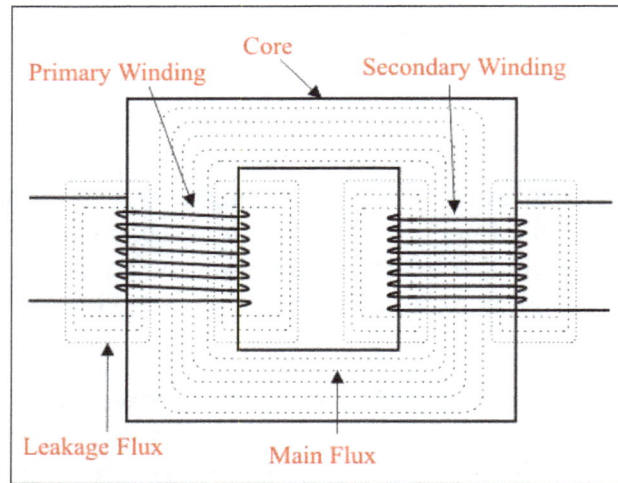

The purpose of the transformer core is to provide a low reluctance path, through which the maximum amount of flux produced by the primary winding is passed through and linked with the secondary winding.

The current that intially passes through the transformer when it is switched on is known as the transformer inrush current.

EMF Equation of the Transformer

As the magnetic flux varies sinusoidally, $\Phi = \Phi_{max} \sin\omega t$, then the basic relationship between induced emf, (E) in a coil winding of N turns is given by:

emf = turns x rate of change,

$$E = N\frac{d\Phi}{dt}$$

$$E = N \times \omega \times \Phi_{max} \times \cos(\omega t)$$

$$E_{max} = N\omega\Phi_{max}$$

$$E_{rms} = \frac{N\omega}{\sqrt{2}} \times \Phi_{max} = \frac{2\pi}{\sqrt{2}} \times f \times N \times \Phi_{max}$$

$$\therefore E_{rms} = 4.44 f N\Phi_{max}$$

Where,

- f - Is the flux frequency in Hertz, $= \omega/2\pi$.

- N - Is the number of coil windings.

- Φ - Is the amount of flux in webers.

This is known as the Transformer EMF Equation. For the primary winding emf, N will be the number of primary turns, (N_p) and for the secondary winding emf, N will be the number of secondary turns, (N_s).

Electrical Power in a Transformer

Another one of the transformer basics parameters is its power rating. The power rating of a transformer is obtained by simply multiplying the current by the voltage to obtain a rating in Volt-amperes, (VA). Small single phase transformers may be rated in volt-amperes only, but much larger power transformers are rated in units of Kilo volt-amperes, (kVA) where 1 kilo volt-ampere is equal to 1,000 volt-amperes, and units of Mega volt-amperes, (MVA) where 1 mega volt-ampere is equal to 1 million volt-amperes.

In an ideal transformer (ignoring any losses), the power available in the secondary winding will be the same as the power in the primary winding, they are constant wattage devices and do not change the power only the voltage to current ratio. Thus, in an ideal transformer the Power Ratio is equal to one (unity) as the voltage, V multiplied by the current, I will remain constant.

That is the electric power at one voltage/current level on the primary is "transformed" into electric power, at the same frequency, to the same voltage/current level on the secondary side. Although the transformer cans step-up (or step-down) voltage, it cannot step-up power. Thus, when a transformer steps-up a voltage, it steps-down the current and vice-versa, so that the output power is always at the same value as the input power. Then we can say that primary power equals secondary power, ($P_p = P_s$).

Power in a Transformer

$$Power_{\text{Primary}} = Power_{\text{Secondary}}$$

$$P_{(PRIM)} = P_{(SEC)} = V_P I_P \cos\theta = V_S I_S \cos\theta_S$$

Where, Φ_p is the primary phase angle and Φ_s is the secondary phase angle.

Note that since power loss is proportional to the square of the current being transmitted, that is: I^2R, increasing the voltage, let's say doubling (×2) the voltage would decrease the current by the same amount, (÷2) while delivering the same amount of power to the load and therefore reducing losses by factor of 4. If the voltage was increased by a factor of 10, the current would decrease by the same factor reducing overall losses by factor of 100.

Transformer Efficiency

A transformer does not require any moving parts to transfer energy. This means that there are no friction or windage losses associated with other electrical machines. However, transformers do suffer from other types of losses called "copper losses" and "iron losses" but generally these are quite small.

A copper loss, also known as I2R loss is the electrical power which is lost in heat as a result of circulating the currents around the transformers copper windings, hence the name. A copper loss represents the greatest loss in the operation of a transformer. The actual watts of power lost can be determined (in each winding) by squaring the amperes and multiplying by the resistance in ohms of the winding (I^2R).

Iron losses, also known as hysteresis is the lagging of the magnetic molecules within the core, in response to the alternating magnetic flux. This lagging (or out-of-phase) condition is due to the fact that it requires power to reverse magnetic molecules; they do not reverse until the flux has attained sufficient force to reverse them.

Their reversal results in friction and friction produce heat in the core which is a form of power loss. Hysteresis within the transformer can be reduced by making the core from special steel alloys.

The intensity of power loss in a transformer determines its efficiency. The efficiency of a transformer is reflected in power (wattage) loss between the primary (input) and secondary (output) windings. Then the resulting efficiency of a transformer is equal to the ratio of the power output of the secondary winding, P_S to the power input of the primary winding, P_P and is therefore high.

An ideal transformer is 100% efficient because it delivers all the energy it receives. Real transformers on the other hand are not 100% efficient and at full load, the efficiency of a transformer is between 94% to 96% which is quite good. For a transformer operating with a constant voltage and frequency with a very high capacity, the efficiency may be as high as 98%. The efficiency, η of a transformer is given as:

Transformer Efficiency

$$efficiency, \eta = \frac{\text{Output Power}}{\text{Input Power}} \times 100\%$$

$$= \frac{\text{Input Power - Losses}}{\text{Input Power}} \times 100\%$$

$$= 1 - \frac{\text{Losses}}{\text{Input Power}} \times 100\%$$

Where, Input, Output and Losses are all expressed in units of power.

Generally when dealing with transformers, the primary watts are called "volt-amps", VA to differentiate them from the secondary watts. Then the efficiency equation above can be modified to:

$$Efficiency, \eta = \frac{\text{Seconday Watts (Output)}}{\text{Primary VA (Input)}}$$

It is sometimes easier to remember the relationship between the transformers input, output and efficiency by using pictures. Here the three quantities of VA, W and η have been superimposed into a triangle giving power in watts at the top with volt-amps and efficiency at the bottom. This arrangement represents the actual position of each quantity in the efficiency formulas.

Transformer Efficiency Triangle

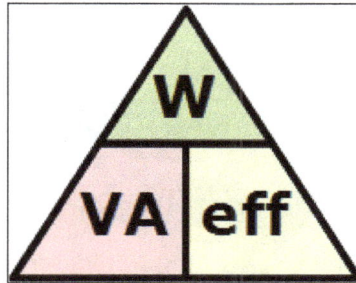

Transposing the above triangle quantities gives us the following combinations of the same equation:

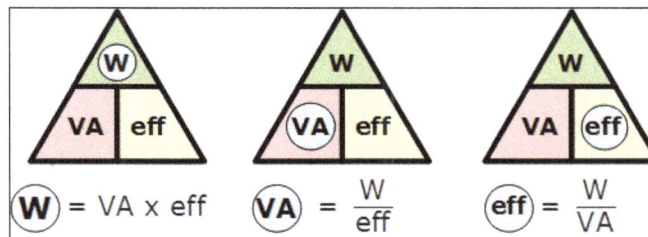

$$\text{W} = \text{VA} \times \text{eff} \qquad \text{VA} = \frac{W}{eff} \qquad \text{eff} = \frac{W}{VA}$$

Then, to find Watts (output) = VA x eff., or to find VA (input) = W/eff., or to find

Efficiency, eff. = W/VA, etc.

Different Types of Transformers

Different types of transformers can be classified based on different criteria like function, core, etc.

Classification according to function:

Step-up Transformer

A step up transformer is the one in which the primary voltage of the coil is lesser than secondary voltage. A Step-up transformer can be used for increasing voltage in the circuit. It is used in flexible ac transmission systems or FACTS by SVC.

Step-down Transformer

A step-down transformer is used for reducing the voltage. The type of transformer in which the primary voltage of the coil is greater than the secondary voltage is termed as step down transformer. Most power supplies use a step-down transformer to reduce the dangerously high voltage to a safer low voltage.

The ratio of the number of turns on each coil, called the turn's ratio determines the ratio of the voltages. A step-down transformer has a large number of turns on its primary (input) coil which is connected to the high voltage mains supply, and a small number of turns on its secondary (output) coil to give a low output voltage.

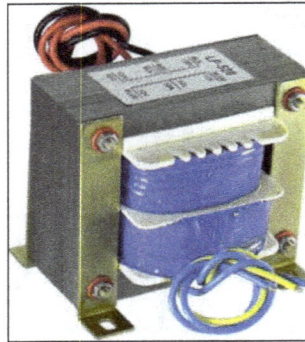

Step down transformer.

TURNS RATIO = (Vp / Vs) = (Np / Ns) Where, Vp = primary (input) voltage Vs = secondary (output) voltage Np = number of turns on primary coil Ns = number of turns on secondary coil Ip = primary (input) current Is = secondary (output) current.

Classification according to core:

1. Core type

2. Shell type

Core Type Transformer

In this type of transformer, the windings are given to the considerable part of the circuit in the core type of the transformer. The coils used are of form-wound and cylindrical type on the core type. It has a single magnetic circuit.

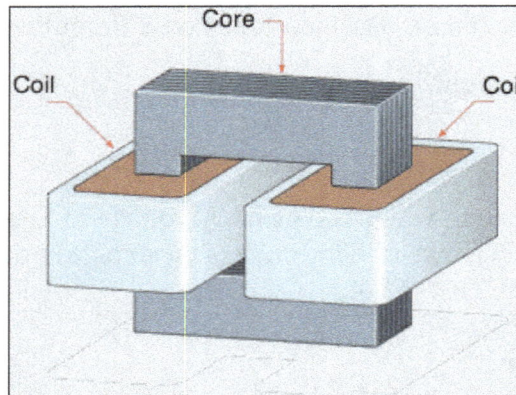

Core Type Transformer.

In core type transformer, the coils are wounded in helical layers with different layers insulated from each other by materials like mica. The core is having two rectangular limbs and the coils are placed on both the limbs in the core type.

Shell Type Transformer

Shell type transformers are the most popular and efficient type of transformers. The shell type transformer has a double magnetic circuit. The core has three limbs and both the winding are placed on the central limbs. The core encircles most parts of the winding. Generally multi-layer disc and sandwich coils are used in shell type.

Shell type transformer.

Each high voltage coil is in between two low voltage coils and low voltage coils are nearest to top and bottom of the yokes. The shell type construction is mostly preferred for operating at very high voltage of transformer.

Natural cooling does not exist in the shell type transformer as the winding in the shell type is surrounded by the core itself. A large number of winding are needed to be removed for better maintenance.

Other Types of Transformers

The types of transformers differ in the manner in which the primary and secondary coils are provided around the laminated steel core of the transformer:

- Based on winding, the transformer can be of three types:
  - Two winding transformer (ordinary type).
  - Single winding (auto type).
  - Three winding (power transformer).
- Based on the arrangement of the coils the transformers are classified as:
  - Cylindrical type.
  - Disc type.
- According to use:
  - Power transformer.
  - Distribution transformer.
  - Instrument transformer.

- Instrument transformer can subdivided into two types:
  - Current transformer.
  - Potential transformer.
- According to the type of cooling the transformer can be of two types:
  - Natural cooling.
  - Oil immersed natural cooled.
  - Oil immersed natural cooled with forced oil circulation.

References

- What-is-electrical-machine: electricaleasy.com, Retrieved 10 February, 2019
- Electric-machines: electrical4u.com, Retrieved 28 April, 2019
- Working-of-generators: elprocus.com, Retrieved 8 January, 2019
- AC-generator-alternator-construction-working: electricaleasy.com, Retrieved 22 March, 2019
- A-c-generator: miniphysics.com, Retrieved 20 January, 2019
- Construction-of-dc-generator: circuitglobe.com, Retrieved 30 May, 2019
- Basic-construction-and-working-of-dc: electricaleasy.com, Retrieved 2 April, 2019
- What-is-a-dc-generator-construction-working-principle-and-applications: elprocus.com, Retrieved 12 August, 2019
- Electrical-motor-types-classification-and-history-of-motor: electrical4u.com, Retrieved 10 June, 2019
- Induction-motor-types-of-induction-motor: electrical4u.com, Retrieved 14 March, 2019
- Dc-motor-basics-types-application: elprocus.com, Retrieved 4 January, 2019
- Dc-motor-or-direct-current-motor: electrical4u.com, Retrieved 21 May, 2019
- Dc-motor-basics-types-application: elprocus.com, Retrieved 25 August, 2019
- What-is-a-transformer: circuitglobe.com, Retrieved 5 July, 2019
- What-is-transformer-definition-working-principle-of-transformer: electrical4u.com, Retrieved 28 February, 2019
- Transformer-basics, Transformer: electronics-tutorials.ws, Retrieved 3 July, 2019
- Working-procedure-on-how-do-transformers-work: elprocus.com, Retrieved 30 June, 2019

6

Measuring Instruments used in Electrical Engineering

There are numerous instruments which are used to measure different quantities in electrical engineering. A few of the commonly used instruments are moving coil galvanometer, ammeter, voltmeter and multimeter. This chapter discusses in detail these instruments related to electrical engineering as well as their applications.

Moving Coil Galvanometer

Moving coil galvanometer is an electromagnetic device that can measure small values of current. It consists of permanent horseshoe magnets, coil, soft iron core, pivoted spring, non-metallic frame, scale, and pointer.

Principle

Torque acts on a current carrying coil suspended in the uniform magnetic field. Due to this, the coil rotates. Hence, the deflection in the coil of a moving coil galvanometer is directly proportional to the current flowing in the coil.

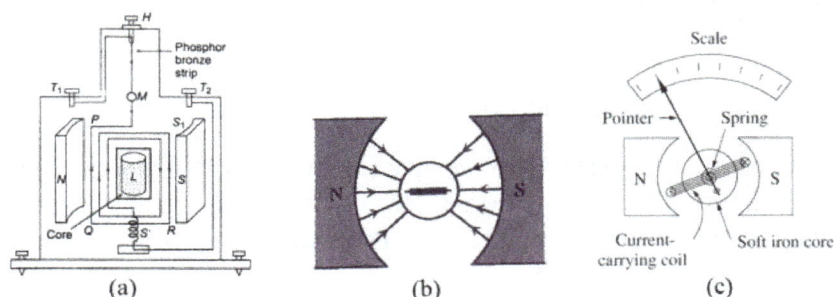

The Moving Coil Galvanometer.

Construction

It consists of a rectangular coil of a large number of turns of thinly insulated copper wire wound over a light metallic frame. The coil is suspended between the pole pieces of a horseshoe magnet by a fine phosphor – bronze strip from a movable torsion head. The lower end of the coil is connected to a hairspring of phosphor bronze having only a few turns.

The other end of the spring is connected to a binding screw. A soft iron cylinder is placed symmetrically inside the coil. The hemispherical magnetic poles produce a radial magnetic field in which the plane of the coil is parallel to the magnetic field in all its positions. A small plane mirror attached to the suspension wire is used along with a lamp and scale arrangement to measure the deflection of the coil.

Working

Let PQRS be a single turn of the coil. A current I flows through the coil. In a radial magnetic field, the plane of the coil is always parallel to the magnetic field. Hence the sides QR and SP are always parallel to the field. So, they do not experience any force. The sides PQ and RS are always perpendicular to the field.

PQ = RS = l, length of the coil and PS = QR = b, breadth of the coil. Force on PQ, F = BI (PQ) = BI*l*. According to Fleming's left-hand rule, this force is normal to the plane of the coil and acts outwards.

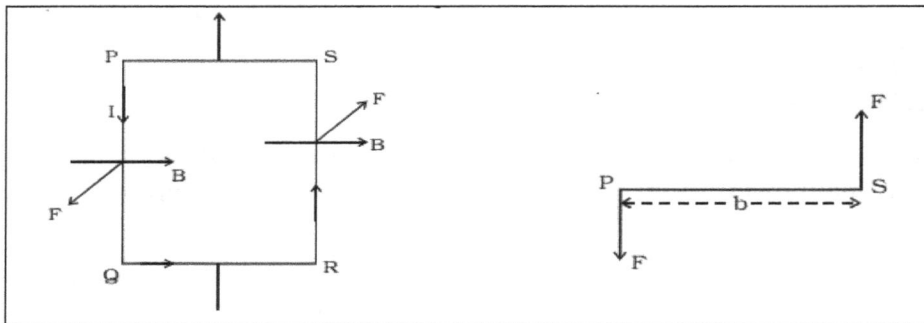

Torque on the coil.

Force on RS, F = BI (RS) = BI*l*. This force is normal to the plane of the coil and acts inwards. These two equal, oppositely directed parallel forces having different lines of action constitute a couple and deflect the coil. If there are *n* turns in the coil, the moment of the deflecting couple = *n* BI*l* – b.

Hence the moment of the deflecting couple = *n*BIA.

When the coil deflects, the suspension wire is twisted. On account of elasticity, a restoring couple is set up in the wire. This couple is proportional to the twist. If θ is the angular twist, then, the moment of the restoring couple = Cθ, where C is the restoring couple per unit twist. At equilibrium, deflecting couple = restoring couple nBIA = Cθ.

Hence we can write, nBIA = Cθ.

$$I = (C / nBA) \times \theta$$

Where, C is the torsional constant of the spring; i.e. the restoring torque per unit twist. The deflection θ is indicated on the scale by a pointer attached to the spring.

The Sensitivity of Moving Coil Galvanometer

The sensitivity of a Moving Coil Galvanometer is defined as the ratio of the change in deflection of the galvanometer to the change in current. Therefore we write, Sensitivity = dθ/di. If a galvanometer gives a larger deflection for a small current it is said to be sensitive. The current in Moving Coil galvanometer is: $I = (C/nBA) \times \theta$.

Therefore, $\theta = (nBA/C) \times I$. Differentiating on both sides wrt I, we have: $d\theta/di = (nBA/C)$. The sensitivity of Moving Coil Galvanometer increases by:

- Increasing the no. of turns and the area of the coil,

- Increasing the magnetic induction,

- Decreasing the couple per unit twist of the suspension fibre.

Applications of Galvanometer

The galvanometer has following applications. They are:

- It is used for detecting the direction of current flows in the circuit. It also determines the null point of the circuit. The null point means the situation in which no current flows through the circuit.

- It is used for measuring the current.

- The voltage between any two points of the circuit is also determined through galvanometer.

Conversion of Galvanometer into an Ammeter

The galvanometer is used as an ammeter by connecting the low resistance wire in parallel with the galvanometer. The potential difference between the voltage and the shunt resistance are equal.

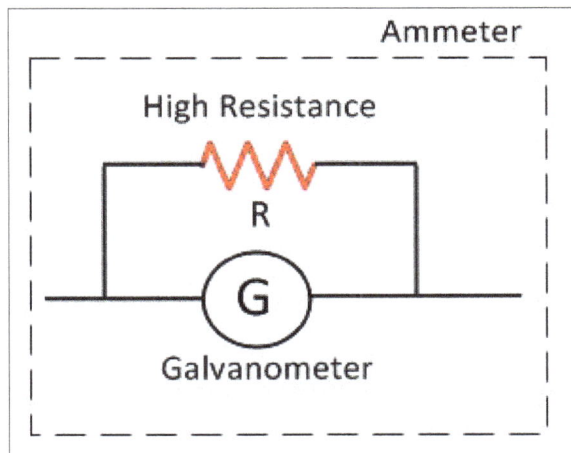

Voltage across galvanometer V_g = Galvanometer Resistance G × Galvanometer Current I_g

Voltage across shunt Resistance $V_s = SI_s$

Where, S = shunt resistance and I_s = current across the shunt.

As the galvanometer and the shunt resistance are connected in potential with the circuit, their potentials are equal.

$$I_s = I - I_s$$
$$V_g = V_s$$
$$GI_g = SI_s$$
$$GI_g = S(I - I_s)$$

Thus, the shunt resistance is given as,

$$S = \frac{GI_g}{(I - I_s)}$$

The value of the shunt current is very small as compared to the supply current.

Conversion of Galvanometer into a Voltmeter

The galvanometer is used as a voltmeter by connecting the high resistance in series with the circuit.

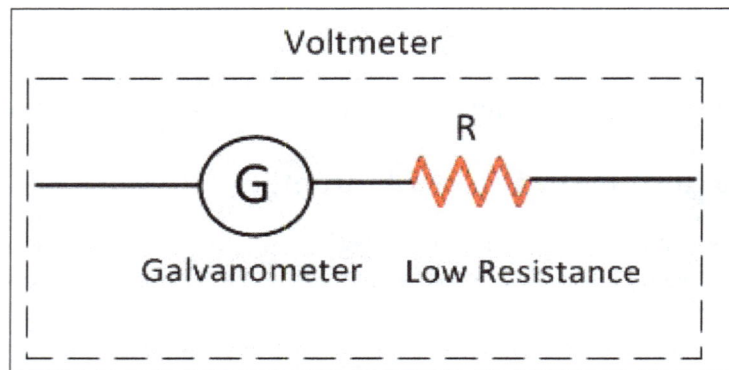

The current passing through the galvanometer is:

$$I_g = \frac{\text{Voltage V}}{\text{Series resistance R}_s}$$

$$R_S = \frac{V}{I_g} - G$$

The range of the voltmeter depends on the value of the resistance connected in series with the circuit.

Advantages and Disadvantages

Advantages:

- Sensitivity increases as the value of n, B, A increases and value of k decreases.

- The eddy currents produced in the frame bring the coil to rest quickly, due to the coil wound over the metallic frame.

Disadvantages:

- Its sensitivity cannot be changed at will.

- Overloading can damage any type of galvanometer.

Ammeter

The meter used for measuring the current is known as the ammeter. The current is the flow of electrons whose unit is ampere. Hence the instrument which measures the flows of current in ampere is known as ampere meter or ammeter.

The ideal ammeter has zero internal resistance. But practically the ammeter has small internal resistance. The measuring range of the ammeter depends on the value of resistance.

Symbolic Representation

The capital alphabet A represents the ammeter in the circuit.

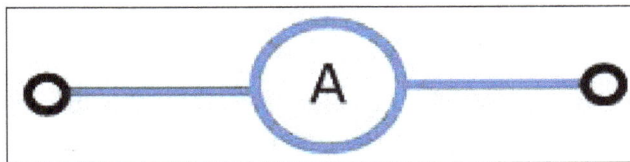

Connection of Ammeter in Circuit

The ammeter is connected in series with the circuit so that the whole electron of measurand current passes through the ammeter. The power loss occurs in ammeter because of the measurand current and their internal resistance. The ammeter circuit has low resistance so that the small voltage drop occurs in the circuit.

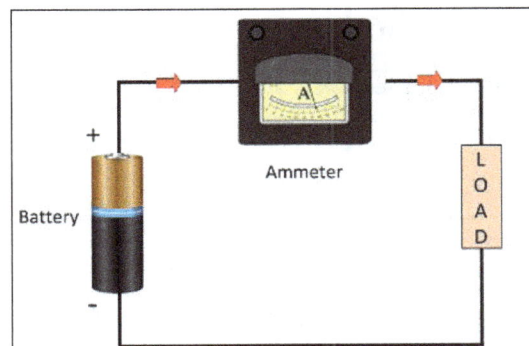

The resistance of the ammeter is kept low because of the two reasons.

- The whole measurand current passes through the ammeter.

- The low voltage drop occurs across the ammeter.

Ammeter Shunt

The high-value current directly passes through the ammeter which damages their internal circuit. For removing this problem, the shunt resistance is connected in parallel with the ammeter.

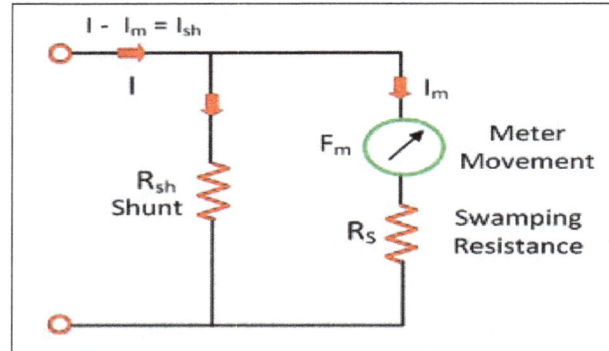

Meter shunt and Swamping Resistance.

If the large measurand current passes through the circuit, the major portion of the current passes through the shunt resistance. The shunt resistance will not affect the working of the ammeter, i.e., the movement of the coil remains same.

Effect of Temperature in Ammeter

The ammeter is a sensitive device which is easily affected by the surrounding temperature. The variation in temperature causes the error in the reading. This can reduce by swamping resistance. The resistance having zero temperature coefficients is known as the swamping resistance. It connects in series with the ammeter. The swamping resistance reduces the effect of temperature on the meter.

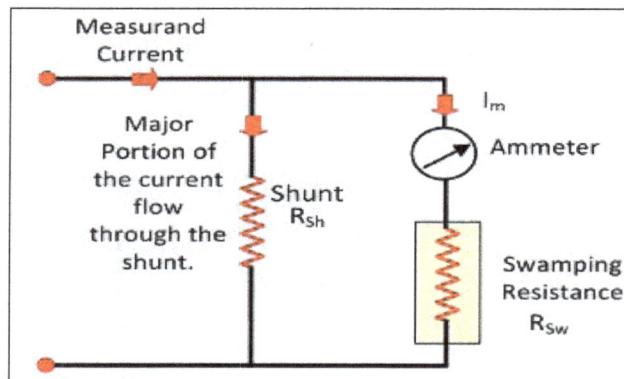

Swamping Resistance.

The ammeter has the inbuilt fuse which protects the ammeter from the heavy current. If substantial current flows through the ammeter, the fuse will break. The ammeter is not able to measure the current until the new one does not replace the fuse.

Working Principle of Ammeter

The main principle of ammeter is that it must have a very low resistance and also inductive reactance. Can't we connect an ammeter in parallel? The answer to this question is that it has very low impedance because it must have very low amount of voltage drop across it and must be connected in series connection because current is same in the series circuit.

Also due to very low impedance the power loss will be low and if it is connected in parallel it becomes almost a short circuited path and all the current will flow through ammeter as a result of high current the instrument may burn. So due to this reason it must be connected in series. For an ideal ammeter, it must have zero impedance so that it has zero voltage drop across it so the power loss in the instrument is zero. But the ideal is not achievable practically.

Classification or Types of Ammeter

Depending on the constructing principle, there are many types of ammeter we get, they are mainly:

1. Permanent Magnet Moving Coil (PMMC) ammeter.

2. Moving Iron (MI) Ammeter.

3. Electrodynamometer type Ammeter.

4. Rectifier type Ammeter.

Depending on these types of measurement we do, we have:

1. DC Ammeter

2. AC Ammeter

DC Ammeter are mainly PMMC instruments, MI can measure both AC and DC currents, also Electrodynamometer type thermal instrument can measure DC and AC, induction meters are not generally used for ammeter construction due to their higher cost, inaccuracy in measurement.

Different Types of Ammeters

PMMC Ammeter

Principle PMMC Ammeter:

When current carrying conductor placed in a magnetic field, a mechanical force acts on the conductor, if it is attached to a moving system, with the coil movement, the pointer moves over the scale.

Explanation: As the name suggests it has permanent magnets which are employed in this kind of measuring instruments. It is particularly suited for DC measurement because here deflection is proportional to the current and hence if current direction is reversed, deflection of the pointer will also be reversed so it is used only for DC measurement. This type of instrument is called D Arnsonval type instrument. It has major advantage of having linear scale, low power consumption, high accuracy. Major disadvantage of being measured only DC quantity, higher cost etc.

Deflecting torque,

$$T = BiNlbNm$$

Where,

- B = Flux density in Wb/m².

- i = Current flowing through the coil in Amp.

- l = Length of the coil in m.

- b = Breadth of the coil in m.

- N = No of turns in the coil.

Extension of Range in a PMMC Ammeter:

Now it looks quite extraordinary that we can extend the range of measurement in this type of instrument. Many of us will think that we must buy a new ammeter to measure higher amount of current and also many of us may think we have to change the constructional feature so that we can measure higher currents, but there is nothing like that, we just have to connect a shunt resistance in parallel and the range of that instrument can be extended, this is a simple solution provided by the instrument.

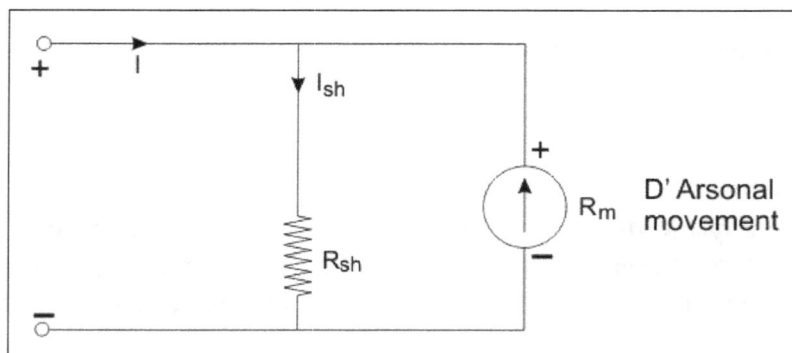

In the figure,

- I = total current flowing in the circuit in Amp.

- I_{sh} is the current through the shunt resistor in Amp.

- R_m is the ammeter resistance in Ohm.

Then,

$$R_{sh} = \frac{R_m}{\dfrac{I}{I - I_{sh}} - 1}$$

MI Ammeter

It is a moving iron instrument, used for both AC and DC, It can be used for both because the deflection θ proportional square of the current so whatever is the direction of current, it shows directional deflection, further they are classified in two more ways;

- Attraction type

- Repulsion type

Its torque equation is:

$$T = \frac{1}{2} I^2 \frac{dL}{d\theta}$$

Where,

- I is the total current flowing in the circuit in Amp.

- L is the self inductance of the coil in Henry.

- θ is the deflection in Radian.

  ○ Attraction Type MI Instrument Principle: When an unmagnetised soft iron is placed in the magnetic field, it is attracted towards the coil, if a moving system attached and current is passed through a coil, it creates a magnetic field which attracts iron piece and creates deflecting torque as a result of which pointer moves over the scale.

  ○ Repulsion Type MI Instrument Principle: When two iron pieces are magnetized with same polarity by passing a current than repulsion between them occurs and that repulsion produces a deflecting torque due to which the pointer moves.The advantages of MI instruments are they can measure both AC and DC, cheap, low friction errors, robustness etc. It is mainly used in AC measurement because in DC measurement error will be more due to hysteresis.

Electrodynamometer Type Ammeter

This can be used to measure both i.e. AC and DC currents. Now we see that we have PMMC and MI instrument for the measurement of AC and DC currents, a question may arise – "why do we need Electrodynamometer Ammeter? If we can measure current accurately by other instrument also?". The answer is Electrodynamometer instruments have the same calibration for both AC and DC i.e. if it is calibrated with DC, then also without calibrating we can measure AC.

Principle Electrodynamometer Type Ammeter:

There we have two coils, namely fixed and moving coils. If a current is passed through two coils it will stay in the zero position due to the development of equal and opposite torque. If somehow, the direction of one torque is reversed as the current in the coil reverses, an unidirectional torque is produced.

For ammeter, the connection is a series one and φ = 0.

Where, φ is the phase angle.

$$T = I^2 \frac{dM}{d\theta}$$

Where,

- I is the amount of current flowing in the circuit in Amp.

- M = Mutual inductance of the coil.

They have no hysteresis error, used for both AC and DC measurement, the main disadvantages are they have low torque/weight ratio, high friction loss, expensive than other measuring instruments etc.

Rectifier Ammeter

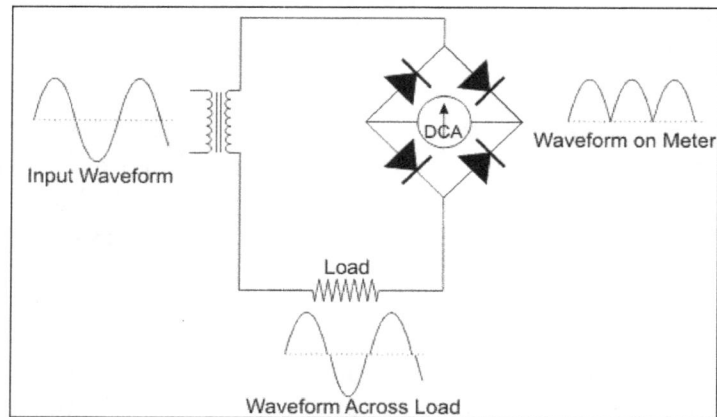

Principle of Rectifier Ammeter:

They are used for AC measurement which is connected to secondary of a current transformer, the secondary current is much less than primary and connected with a bridge rectifier to a moving coil ammeter.

Advantages:

- It can be used in high frequency also.

- Uniform scale for most of the ranges.

Voltmeter

The instrument which measures the voltage or potential difference in volts is known as the voltmeter. It works on the principle that the torque is generated by the current which induces because of measurand voltage and this torque deflects the pointer of the instrument. The deflection of the pointer is directly proportional to the potential difference between the points. The voltmeter is always connected in parallel with the circuit.

Symbolic Representation of Voltmeter

The voltmeter is represented by the alphabet V inside the circle along with the two terminals.

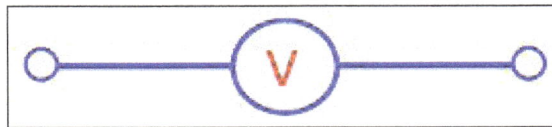

Working Principle of Voltmeter

The main principle of voltmeter is that it must be connected in parallel in which we want to measure the voltage. Parallel connection is used because a voltmeter is constructed in such a way that it has a very high value of resistance. So if that high resistance is connected in series than the current flow will be almost zero which means the circuit has become open.

If it is connected in parallel, than the load impedance comes parallel with the high resistance of the voltmeter and hence the combination will give almost the same the impedance that the load had. Also in parallel circuit we know that the voltage is same so the voltage between the voltmeter and the load is almost same and hence voltmeter measures the voltage.

For an ideal voltmeter, we have the resistance is to be infinity and hence the current drawn to be zero so there will be no power loss in the instrument. But this is not achievable practically as we cannot have a material which has infinite resistance.

Classification or Types of Voltmeter

According to the construction principle, we have different types of voltmeters, they are mainly:

1. Parmanent Magnet Moving coil (PMMC) Voltmeter.

2. Moving Iron (MI) Voltmeter.

3. Electro Dynamometer Type Voltmeter.

4. Rectifier Type Voltmeter

5. Induction Type Voltmeter.

6. Electrostatic Type Voltmeter.

7. Digital Voltmeter (DVM).

Depending on these types of measurement we do, we have:

- DC Voltmeter

- AC Voltmeter

For DC voltmeters PMMC instruments are used, MI instrument can measure both AC and DC voltages, electrodynamometer type, thermal instrument can measure DC and AC voltages as well. Induction meters are not used because of their high cost, inaccuracy in measurement. Rectifier type voltmeter, electrostatic type and also digital voltmeter (DVM) can measure both AC and DC voltages.

PMMC Voltmeter

When current carrying conductor placed in a magnetic field, a mechanical force acts on the conductor, if it is attached to a moving system, with the coil movement, the pointer moves over the scale.

PMMC instruments have parmanent magnets. It is suited for DC measurement because here deflection is proportional to the voltage because resistance is constant for a material of the meter and hence if voltage polarity is reversed, deflection of the pointer will also be reversed so it is used only for DC measurement. This type of instrument is called D'Arnsonval type instrument. It has advantages of having linear scale, power consumption is low, high accuracy.

Major disadvantages are:

- It only measures DC quantity, higher cost etc.

Deflecting torque, $T = BiNIb\,Nm$

Where,

- B = Flux density in Wb/m².

- i = V/R where V is the voltage to be measured and R is the resistance of the load.

- l = Length of the coil in m.

- b = Breadth of the coil in m.

- N = No of turns in the coil.

Extension of Range in a PMMC Voltmeter

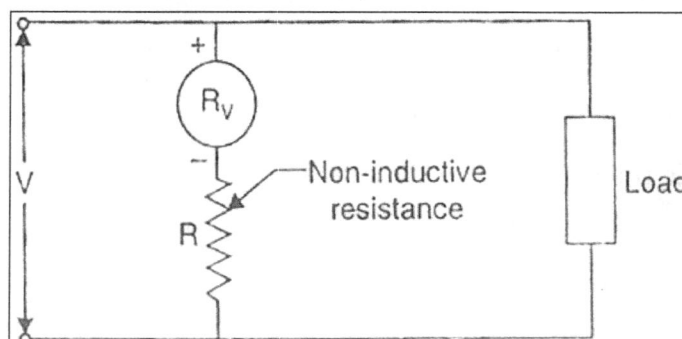

In the PMMC voltmeters we have the facility of extending the range of measurement of voltage also. Just connecting a resistance in series with the meter we can extend the range of measurement. Let,

- V is the supply voltage in volts.

- R_v is the voltmeter resistance in Ohm.

- R is the external resistance connected in series in ohm.

- V_1 is the voltage across the voltmeter.

Then the external resistance to be connected in series is given by:

$$R = \frac{V - V_1}{V_1} \times R_v$$

MI Voltmeter

MI instruments mean moving iron instrument. It is used for both AC and DC measurements, because the deflection θ proportional square of the voltage assuming impedance of the meter to be constant, so whatever is polarity of the voltage, it shows directional deflection, further they are classified in two more ways,

- Attraction type

- Repulsion type

Its Torque equation is: $T = \frac{1}{2} \times I^2 \frac{dL}{d\theta}$

Where,

- I is the total current flowing in the circuit in Amp. I = V/Z.

- Where, V is the voltage to be measured and Z is the impedance of the load.

- L is the self inductance of the coil in Henry.

- θ is the deflection in Radian.

Attraction Type MI Instrument Principle

If an unmagnetised soft iron is placed in the magnetic field, it is attracted towards the coil, if a pointer is attached to the systems and current is passed through a coil as a result of the applied voltage, it creates a magnetic field which attracts iron piece and creates deflecting torque as a result of which pointer moves over the scale.

Repulsion Type MI Instrument Principle

When two iron pieces are magnetized with the same polarity by passing a current which done by applying a voltage across the voltmeter than repulsion between them occurs and that repulsion produces a deflecting torque due to which the pointer moves.

The advantages are it measure both AC and DC, it is cheap, low friction errors, Robust etc. It is mainly used in AC measurement because in DC measurement error will be more due to hysteresis.

Electrodynamometer Type Voltmeter

Electrodynamometer instruments are used because they have the same calibration for both AC and DC i.e. if it is calibrated with DC, then also without calibrating we can measure AC.

Electrodynamometer Type Voltmeter Principle

We have two coils, fixed and moving coils. If a voltage is applied at the two coils as a result of which current flows two coils it will stay in the zero position due to the development of equal and opposite torque. If the direction of one torque is reversed as the current in the coil reverses, an undirectional torque is produced.

For voltmeter, the connection is a parallel one and both fixed and moving coils are connected in series with non-inductive resistance.

$\varphi = 0$ where φ is the phase angle.

$$T = I^2 \times \frac{dM}{d\theta}$$

Where,

- I is the amount of current flowing in the circuit in Amp = V/Z.

- V and Z are the applied voltages and impedance of the coil respectively.

- M = Mutual inductance of the coil.

They have no hysteresis error, can be used for both AC and DC measurement, the main disadvantages are having low torque/weight ratio, high friction loss, expensive than other instruments etc.

Rectifier Voltmeter

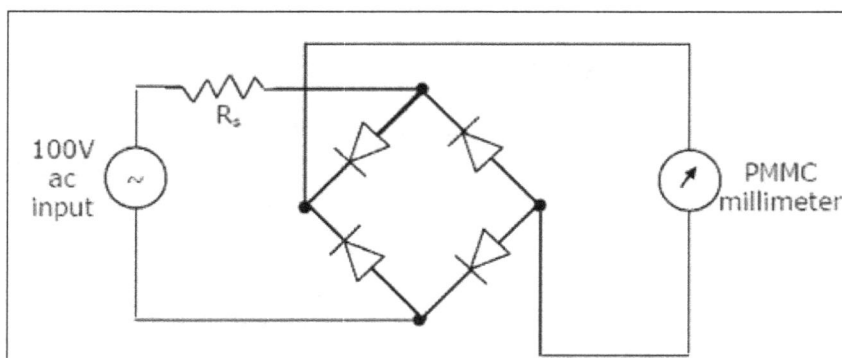

Rectifier Voltmeter Principle

They are used for AC or DC measurements. For DC measurement we have to connect a PMMC meter which measures pulsating DC voltage which measures rectified voltage which is connected across the bridge rectifier.

Advantages of Rectifier Voltmeter

1. Can be used in high frequency.

2. It has uniform scale for most of the ranges.

Disadvantages of having an error due to temperature decrease in sensitivity in AC operation.

Digital Voltmeters (DVM) Principle

The Digital Voltmeter is an instrument which can give the output voltage not by deflection but directly indicating the value. It is a very good instrument to measure the voltage as it eliminates completely the error due to parallax, approximation in measurement, high-speed reading can be done and it can also be stored in memory for further analysis. The main principle is that the value is measured by the same circuit arrangement but that value is not used to deflect the pointer, but it is fed to the analog to digital converter and displayed as the digital value.

Ohmmeter

The ohmmeter is an instrument which measures resistance of a quantity. Resistance in the electrical sense means the opposition offered by a substance to the current flow in the device. Every device has a resistance, it may be large or small and it increases with temperature for conductors, however for semiconducting devices the reverse is true.

There are many types of ohmmeters available such as:

1. Series ohmmeter.

2. Shunt ohmmeter.

3. Multi range ohmmeter.

Working Principle of Ohmmeter

The instrument is connected with a battery, a series adjustable resistor and an instrument which gives the reading. The resistance to be measured is connected at terminal ob. When the circuit is completed by connecting output resistance, the circuit current flows and so the deflection is measured.

When the resistance to be measured is very high then current in the circuit will be very small and the reading of that instrument is assumed to be maximum resistance to be measured. When resistance to be measured is zero then the instrument reading is set to zero position which gives zero resistance.

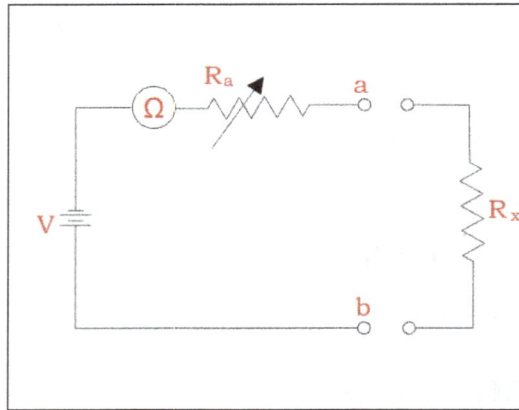

D'arsonval Movement

This type of movement is used in DC measuring instruments. The main principle in these types of instruments is that when a current carrying coil which is placed in a magnetic field, it feels a force and that force can deflect the pointer of a meter and we get the reading in the instrument.

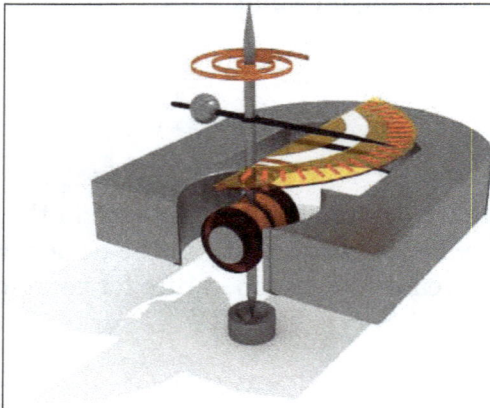

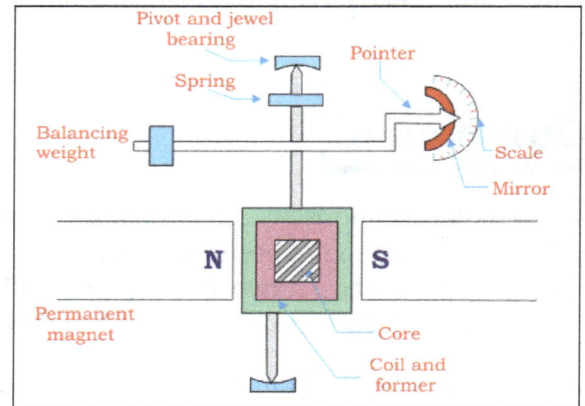

Construction of Dársonval Instrument.

This type of instruments consists of a permanent magnet and a coil which carries current and is placed in between them. The coil may be of rectangular or circular in shape. The iron core is used to provide flux of low reluctance so it produces high intensity magnetic field. Due to high intensity magnetic fields the deflecting torque produced is of large value due to which sensitivity of the meter is also increased. Current which entered comes out of two control springs, one in the upper side and one in the lower side.

If the direction of current is reversed in these types of instruments, then torque direction will also be reversed so these types of instruments are applicable in DC measurements only. The deflecting torque is directly proportional to the deflection angle hence these types of instruments have the linear scale. To limit the deflection of the pointer we have to use damping which provides an equal and opposite force to the deflecting torque and hence the pointer comes to rest at a certain value.

The indication of the breeding is given by a mirror in which a beam of light is reflected on to the scale and hence deflection can be measured.

There are many advantages due to which we use D'Arsonval type instrument. They are-

1. They have uniform scale.

2. Effective eddy current damping.

3. Low power consumption.

4. No hysteresis loss.

5. They are not affected by stray fields.

Owing to possess those major advantages we can use this type of instrument. However they suffer from drawbacks such as:

1. It cannot be used in AC.

2. Costlier compared to MI instruments.

3. There may be error due to ageing of springs by which we may not get accurate result.

However in case of resistance measurement we go for DC measurement because of the advantages offered by PMMC instruments and we multiply that resistance by 1.6 to find out AC resistance, so these instruments are much widely used due to their advantages. The disadvantages offered by it are dominated by the advantages so they are used.

Series Type Ohmmeter

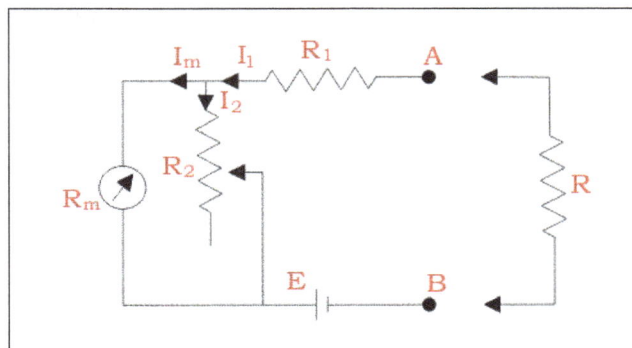

Basic series type ohmmeter.

The series type ohmmeter consists of a current limiting resistor R_1, Zero adjusting resistor R_2, EMF source E, Internal resistance of D'Arsonval movement R_m and the resistance to be measured R.

When there is no resistance to be measured, current drawn by the circuit will be maximum and the meter will show a deflection. By adjusting R_2 the meter is adjusted to a full scale current value since the resistance will be zero at that time. The co-responding pointer indication is marked as zero. Again when the terminal AB is opened it provides very high resistance and hence almost zero current will flow through the circuit. In that case the pointer deflection is zero which is marked at very high value for resistance measurement. So a resistance between zeros to a very high value is marked and hence can be measured. So, when a resistance is to be measured, the current value will be somewhat less than the maximum and the deflection is recorded and accordingly resistance is measured. This method is good but it possess certain limitations such as the decrease in potential of the battery with its use so adjustment must be made for every use. The meter may not read zero when terminals are shorted, these types of problem may arise which is counteracted by the adjustable resistance connected in series with the battery.

Shunt Type Ohmmeter

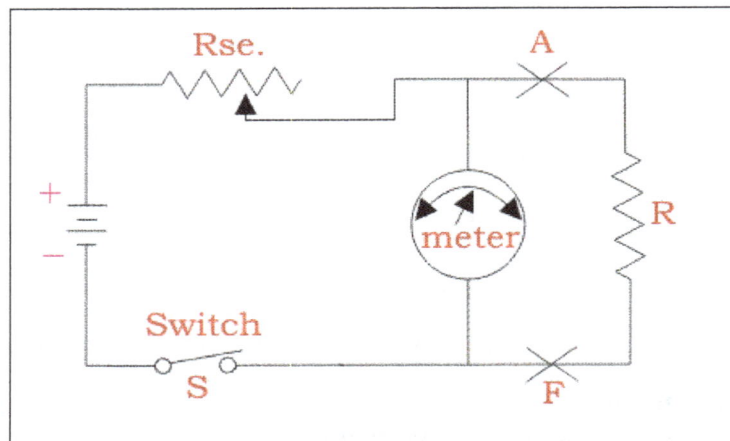

Shunt type ohmmeter.

In this type of meters we have a battery source and an adjustable resistor is connected in series with the source. We have connected the meter in parallel to the resistance which is to be measured. There is a switch by the use of which we can on or off the circuit. The switch is opened when it is not in use. When the resistance to be measured is zero, the terminals A and F are shorted so the current through the meter will be zero. The zero position of the meter denotes the resistance to be zero. When the resistance connected is very high, then a small current will flow the terminal AF and hence full scale current is allowed to flow through the meter by adjusting the series resistance connected with the battery. So, full scale deflection measures very high resistance. When the resistance to be measured is connected between A and F, The pointer shows a deflection by which we can measure the resistance values.

In this case also, the battery problem may arise which can be counteracted by adjusting the resistance. The meter may have some error due to its repeated use also.

Multi Range Ohmmeter

This instrument provides the reading up to a very wide range. In this case we have to select the range switch according to our requirement. An adjuster is provided so that we can adjust the initial reading to be zero. The resistance to be measured is connected in parallel to the meter. The meter is adjusted so that it shows full scale deflection when the terminals in which the resistance connected is full scale range through the range switch. When the resistance is zero or short circuit, there is no current flow through the meter and hence no deflection. Suppose we have to measure a resistance less than 1 ohm, then the range switch is selected at 1 ohm range at first. Then that resistance is connected in parallel and the corresponding meter deflection is noted. For 1 ohm resistance it shows full scale deflection but for the resistance other than 1 ohm it shows a deflection which is less than the full load value and hence resistance can be measured. This is the most suitable method of all the ohmmeters as we can get accurate reading in this type of meters. So this meter is most widely used now days.

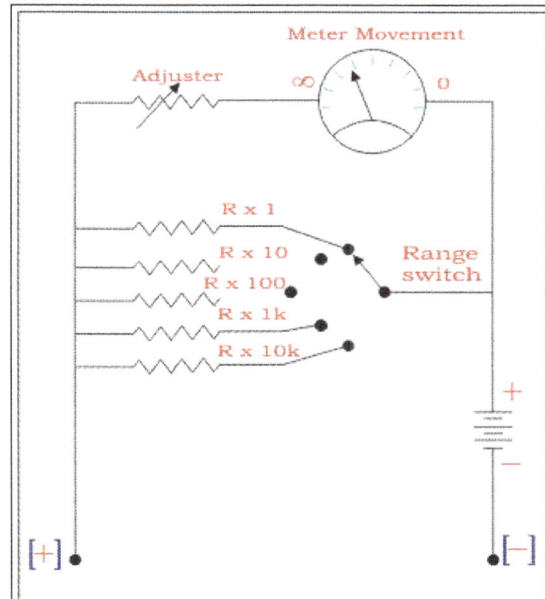

Multirange ohmmeter.

Multimeter

A Multimeter is an electronic instrument, every electronic technician and engineers widely used piece of test equipment. Multimeter is mainly used to measure the three basic electrical characteristics of voltage, current and resistance. It can also be used to test continuity between two points in a electrical circuit. Multimeter has multi functionalities like, it acts like ammeter, voltmeter and ohmmeter. It is a handheld device with positive and negative indicator needle over a numeric LCD digital display. Multimeters can be used for testing batteries, household wiring, electric motors and power supplies.

Types of Multimeters

There are different types of multimeter like Analog, Digital and Fluke multimeters.

Digital Multimeter

We mostly used multimeter is digital multimeter (DMM). The DMM performs all functions from AC to DC other than analog. It has two probes positive and negative indicated with black and red color is shown in figure. The black probe connected to COM JACK and red probe connected by user requirement to measure ohm, volt or amperes. The jack marked VΩ and the COM jack on the right of the picture are used for measuring voltages, resistance and for testing a diode. The two jacks are utilized when LCD display that shows what is being measured (volts, ohms, amps, etc.). Overload protection that prevents damage to the meter and the circuit, and protects the user.

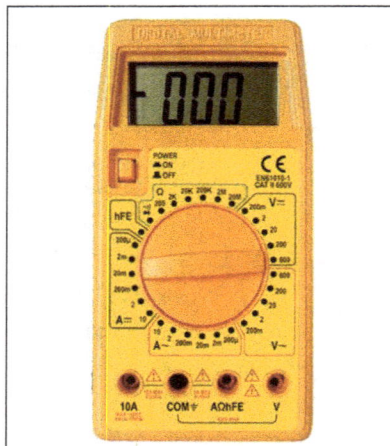

Digital Multimeter.

The Digital Multimeter basically consists of a LCD display, a knob to select various ranges of the three electrical characteristics, an internal circuitry consisting of a signal conditioning circuitry, an analog to digital converter. The PCB consists of concentric rings which are connected or disconnected based on the position of the knob. Thus as the required parameter and the range is selected, the section of the PCB is activated to perform the corresponding measurement. To measure the resistance, current flows from a constant current source through the unknown resistor and the voltage across the resistor is amplified and fed to a Analog to Digital Converter and the resultant output in form of resistance is displayed on the digital display. To measure an unknown AC voltage, the voltage is first attenuated to get the suitable range and then rectified to DC signal and the analog DC signal is fed to A/D converter to get the display, which indicates the RMS value of the AC signal. Similarly to measure an AC or DC current, the unknown input is first converted to voltage signal and then fed to analog to digital converter to get the desired output(with rectification in case of AC signal).

Advantages of a Digital Multimeter are its output display which directly shows the measured value, high accuracy, and ability to read both positive and negative values.

Analog Multimeter

The Analog Multimeter or VOM (Volt-Ohm-Milliammeter) is constructed using a moving coil meter and a pointer to indicate the reading on the scale. The moving coil meter consists of a coil wound around a drum placed between two permanent magnet. As current passes through the coil, magnetic field is induced in the coil which reacts with the magnetic field of the permanent magnets and the resultant force causes the pointer attached to the drum to deflect on the scale, indicating the current reading. It also consists of springs attached to the drum which provides an opposing force to the motion of the drum to control the deflection of the pointer.

For measurement of DC current, the D Arsonval movement described above can be directly used.

However the current to be measured should be lesser than the full scale deflection current of the meter. For higher currents the current divider rule is applied. Using different values of shunt resistors, the meter can also be used for multi range current measurement. For current measurement the instrument is to be connected in series with the unknown current source.

Analog Multimeter.

For measurement of DC voltage, a resistor is connected in series with the meter and the meter resistance is taken into account such that the current passing through the resistor is same as the current passing through the meter and the whole reading indicates the voltage reading. For voltage measurement, the instrument is to be connected in parallel with the unknown voltage source. For multirange measurement, different resistors of different values can be used, which are connected in series with the meter.

For measurement of resistance, the unknown resistance is connected in series with the meter and across a battery, such that the current passing through the meter is directly proportional to the unknown resistance.

For AC voltage or current measurement, the same principle is applied, except for the fact that the AC parameter to be measured is first rectified and filtered to get the DC parameter and the meter indicates the RMS value of the AC signal.

Advantages of an Analog Multimeter are that it is inexpensive, doesn't require a battery, can measure fluctuations in the readings.

The two main factors affecting the measurement are the sensitivity and the accuracy. Sensitivity refers to the reciprocal of the full scale deflection current and is measured in ohms per volt.

Fluke Multimeter

The fluke multimeters are protected against the transient voltage. It is a small portable device used to measure voltage, current and test diodes. The multi meter has multi selectors to select the desired function. The fluke MM automatically ranges to select most measurements. This means the magnitude of the signal does not have to be known or determined to take an accurate reading, it directly moved to the appropriate port for the desired measurement. The fuse is protected to prevent the damage, if connected to wrong port.

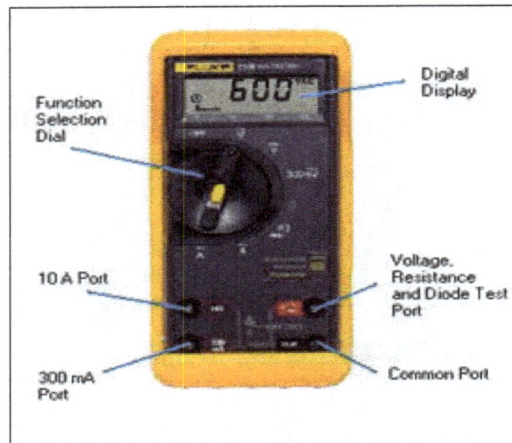

Fluke Multimeter.

Applications

The applications of ammeter mainly involves in various electrical and electronic projects for the purpose of components testing and also used in different measurement applications in multimeter.

- Temperature and Environmental Applications

  ○ Low cost weather station

  ○ DMM internal temperature

- Voltage Measurements

  ○ High and low value DC measurement

  ○ Peak to Peak and DC average measurement

- Current Measurements

  ○ DC current measurement

  ○ True RMS AC current

- Resistance Measurement

  ○ Micro ohm meter

  ○ Measuring resistance with constant voltage

  ○ Measuring resistance with constant current

- Time and Frequency measurement

  ○ Fast frequency

  ○ Time measurement

Energy Meter

The meter which is used for measuring the energy utilises by the electric load is known as the energy meter. The energy is the total power consumed and utilised by the load at a particular interval of time. It is used in domestic and industrial AC circuit for measuring the power consumption. The meter is less expensive and accurate.

Construction of Energy Meter

The construction of the single phase energy meter is shown in the figure.

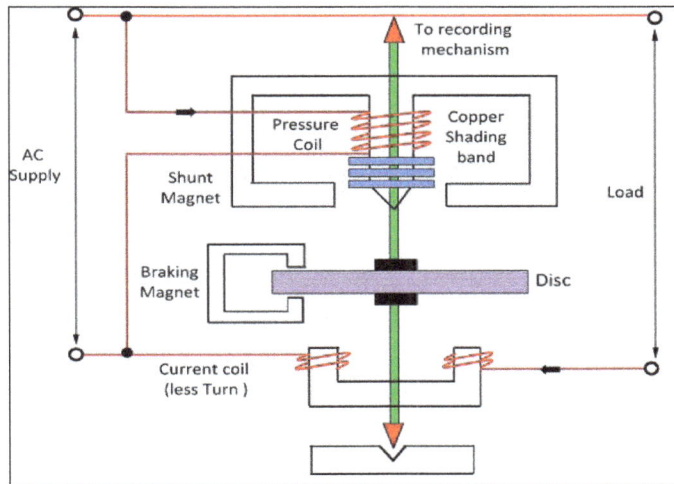

Induction Type Energy Meter.

The energy meter has four main parts. They are the

1. Driving System

2. Moving System

3. Braking System

4. Registering System

1. Driving System: The electromagnet is the main component of the driving system. It is the temporary magnet which is excited by the current flow through their coil. The core of the electromagnet is made up of silicon steel lamination. The driving system has two electromagnets. The upper one is called the shunt electromagnet, and the lower one is called series electromagnet.

The series electromagnet is excited by the load current flow through the current coil. The coil of the shunt electromagnet is directly connected with the supply and hence carry the current proportional to the shunt voltage. This coil is called the pressure coil.

The centre limb of the magnet has the copper band. These bands are adjustable. The main function of the copper band is to align the flux produced by the shunt magnet in such a way that it is exactly perpendicular to the supplied voltage.

2. Moving System: The moving system is the aluminium disc mounted on the shaft of the alloy. The disc is placed in the air gap of the two electromagnets. The eddy current is induced in the disc because of the change of the magnetic field. This eddy current is cut by the magnetic flux. The interaction of the flux and the disc induces the deflecting torque.

When the devices consume power, the aluminium disc starts rotating, and after some number of rotations, the disc displays the unit used by the load. The number of rotations of the disc is counted at particular interval of time. The disc measured the power consumption in kilowatt hours.

3. Braking system: The permanent magnet is used for reducing the rotation of the aluminium disc. The aluminium disc induces the eddy current because of their rotation. The eddy current cut the magnetic flux of the permanent magnet and hence produces the braking torque.

This braking torque opposes the movement of the disc, thus reduces their speed. The permanent magnet is adjustable due to which the braking torque is also adjusted by shifting the magnet to the other radial position.

4. Registration (Counting Mechanism): The main function of the registration or counting mechanism is to record the number of rotations of the aluminium disc. Their rotation is directly proportional to the energy consumed by the loads in the kilowatt hour.

The rotation of the disc is transmitted to the pointers of the different dial for recording the different readings. The reading in kWh is obtained by multiply the number of rotations of the disc with the meter constant. The figure of the dial is shown below.

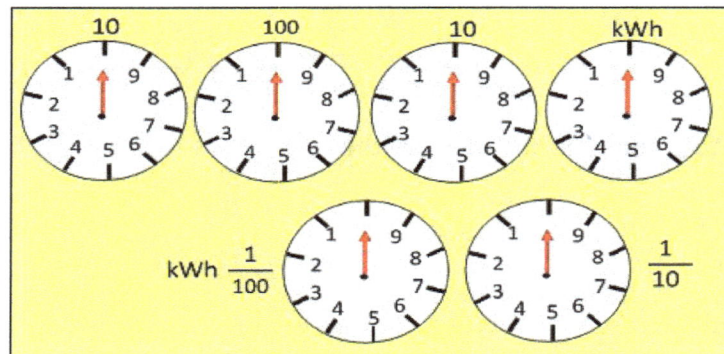

Pointer Type of Register.

Working of the Energy Meter

The energy meter has the aluminium disc whose rotation determines the power consumption of the load. The disc is placed between the air gap of the series and shunt electromagnet. The shunt magnet has the pressure coil, and the series magnet has the current coil.

The pressure coil creates the magnetic field because of the supply voltage, and the current coil produces it because of the current.

The field induces by the voltage coil is lagging by $90°$ on the magnetic field of the current coil because of which eddy current induced in the disc. The interaction of the eddy current and the magnetic field causes torque, which exerts a force on the disc. Thus, the disc starts rotating.

The force on the disc is proportional to the current and voltage of the coil. The permanent magnet controls Their rotation. The permanent magnet opposes the movement of the disc and equalises it on the power consumption. The cyclometer counts the rotation of the disc.

Theory of Energy Meter

The pressure coil has the number of turns which makes it more inductive. The reluctance path of their magnetic circuit is very less because of the small length air gap. The current I_p flows through the pressure coil because of the supply voltage, and it lags by 90º.

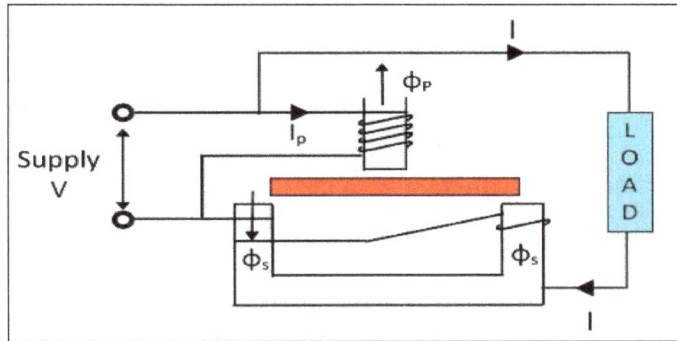

Working of Energy Meter.

The I_p produces the two Φ_p which is again divided into Φ_{p1} and Φ_{p2}. The major portion of the flux Φ_{p1} passes through the side gap because of low reluctance. The flux Φ_{p2} goes through the disc and induces the driving torque which rotates the aluminium disc.

The flux Φ_p is proportional to the applied voltage, and it is lagged by an angle of 90º. The flux is alternating and hence induces an eddy current I_{ep} in the disc.

The load current passes through the current coil induces the flux Φ_s. This flux causes the eddy current I_{es} on the disc. The eddy current I_{es} interacts with the flux Φ_p, and the eddy current I_{ep} interacts with Φ_s to produce another torque. These torques are opposite in direction, and the net torque is the difference between these two.

The phasor diagram of the energy meter is shown in the figure.

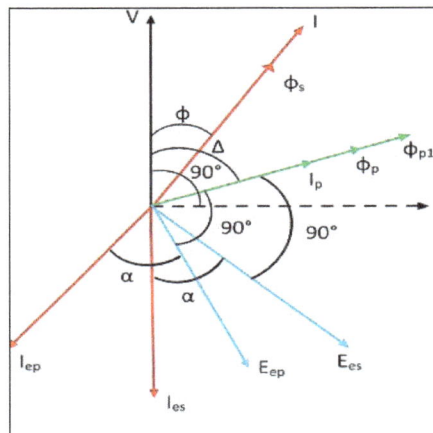

Phasor Diagram of Energy Meter.

Let,

- V – applied voltage,
- I – load current,
- $\emptyset$ – the phase angle of load current,
- I_p – pressure angle of load,
- Δ – the phase angle between supply voltage and pressure coil flux,
- f – frequency,
- Z – impedance of eddy current,
- $\propto$ – the phase angle of eddy current paths,
- E_{ep} – eddy current induced by flux,
- I_{ep} – eddy current due to flux,
- E_{ev} – eddy current due to flux,
- I_{es} – eddy current due to flux.

The net driving torque of the dis is expressed as:

$$T_d \propto \phi_1\phi_2\frac{f}{Z}\sin\beta\cos\alpha = K_1\phi_1\phi_2\frac{f}{Z}\sin\beta\cos\alpha$$

Where,

- K_1 – constant
- Φ_1 and Φ_2 are the phase angle between the fluxes. For energy meter, we take Φ_p and Φ_s.
- β – phase angle between fluxes Φ_p and $\Phi_p = (\Delta - \Phi)$, therefore

$$Dricving\ Torque,\ T_d = K_1\phi_1\phi_2\frac{f}{Z}\sin(\Delta-\phi)\cos\alpha$$

$$But,\ \phi_p \propto V,\ and\ \phi_p \propto I$$

$$T_d \propto K_2 VI\frac{f}{Z}\sin(\Delta-\phi)\cos\alpha$$

If f, Z and α are constants,

$$T_d = K_3 VI\sin(\Delta-\phi)$$

If N is steady speed, braking torque,

$$T_B = K_4 N$$

At steady state, the speed of the driving torque is equal to the braking torque.

$$K_4 N = K_3 VI(\Delta - \phi)$$
$$N = KVI \sin(\Delta - \phi)$$

If $\Delta = 90°$,

Speed,

$$N = KVI \sin(90° - \phi) = KVI \cos\phi$$
$$= K \times power$$

The speed of the rotation is directly proportional to the power.

$$Total\ number\ of\ revolution = \int N dt = K \int VI \sin(\Delta - \phi)$$

If $\Delta = 90°$, total number of revolutions,

$$= K \int VI \cos\phi dt$$
$$= K \int power\ dt = K\ X\ energy$$

The three phase energy meter is used for measuring the large power consumption.

Wattmeter

A wattmeter is an instrument which gives a visual indication of the amount of electrical energy being supplied to a circuit. This indication is expressed in watts which is the standard unit of measure for electrical energy supply or consumption. There are two commonly used types of wattmeter: analog and digital. Analog meters indicate power supply via a needle and scale indicator while digital instruments display the power usage on a liquid crystal display (LCD). Wattmeters are typically rated for a set voltage range but may include features such as coil taps which allow for multiple voltages.

All electrical equipment consumes power subject to a set of known constants which include the rated voltage, the current usage expressed in amps, and the overall energy usage expressed in watts. Some types of electrical appliances or installations use far more energy than others of similar voltage ratings. A wattmeter allows for power usage to be monitored establishing whether circuits are operating correctly. This information is crucial in larger installations where large resistive loads are used. The wattmeters in such installations allow operators and technicians to keep track of individual circuit health and overall power supply balancing and consumption.

Construction of a Wattmeter

The internal construction of a wattmeter is such that it consists of two cols. One of the coil is in series and the other is connected in parallel. The coil that is connected in series with the circuit is known as the current coil and the one that is connected in parallel with the circuit is known as the voltage coil.

These coils are named according to the convention because the current of the circuit passes through the current coil and the voltage is dropped across the potential coil, also named as the voltage coil.

The needle that is supposed to move on the marked scale to indicate the amount of power is also attached to the potential coil. The reason for this is that the potential coil is allowed to move where-as the current coil is kept fixed.

The mechanical construction of a wattmeter is shown in the figure.

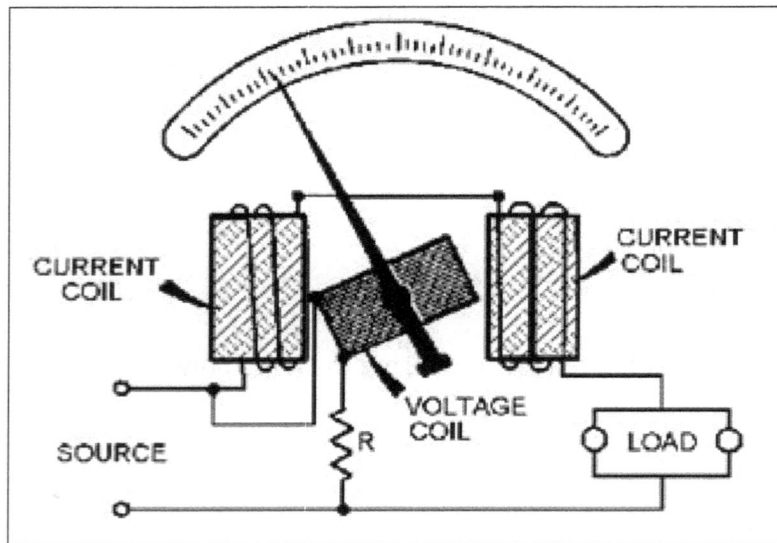

Working of a Wattmeter

When the current passes through the current coil, it creates an electromagnetic field around the coil. The strength of this electromagnetic field is directly proportional to the amount of current passing through it.

In case of DC current, the current is also in phase with its generated electromagnetic field. The voltage is dropped across the potential coil and as a result of this complete process, the needle moves across the scale. The needle deflection is such that it is according to the product of the current passing and the voltage dropped, that is, P = VI.

This was the case of DC power. We know that the AC power is given by the formula P = VIcosθ, and we know that this cosθ factor is because of the fact that the current and voltage are not in phase.

But the question that arises here is that how will a wattmeter measure the AC power and this power factor? So the wattmeter simply measures the average power in case if AC power is required.

The measurement principle of wattmeter is shown in the figure.

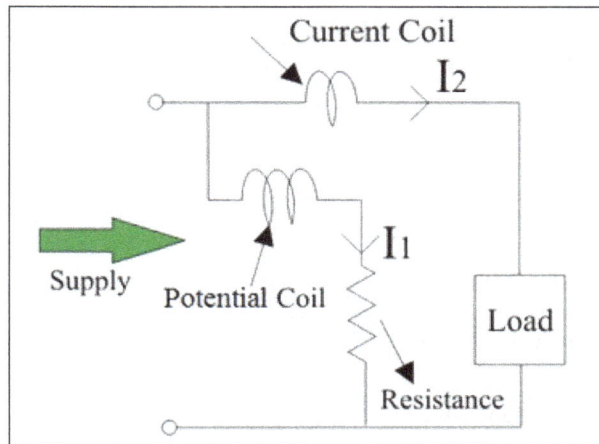

Applications of Wattmeter

1. As other measuring instruments, watt meters are also used extensively in electrical circuit measurement and debugging.

2. They are also used in industries to check the power rating and consumption of electrical appliances.

3. Electromagnetic watt meters are used to measure utility frequencies.

4. They are used with refrigerators, electric heaters and other equipment to measure their power ratings.

Three Phase Wattmeter

Three-Phase Wattmeter is used for measuring the power of the three-phase circuit. In three-phase Wattmeter, the two separate Wattmeter are mounted together in the single unit. Their moving coils are placed on the same spindle.

Three-Phase Wattmeter has two elements. The single element is the combination of the pressure coil and the current coil. The current coils are considered as the fixed coil, and the pressure coils are the moving coil of the Wattmeter.

Working Principle of Three Phase Wattmeter

It works on the principle that the torque develops on the current carrying conductor when it is placed in the magnetic field. The measurand power when passes through the moving coils, the torque develops on the coil. The torque is the type of mechanical force whose effect can deflect the object in circular motion.

In three-phase Wattmeter, the torque develops on both the elements. The value of torque on each element is proportional to the power passes through it. The total torque on the three-phase Watt-meter is the sum of the torque on individual Wattmeter.

Let understand this with the help of the mathematical expressions.

Consider the deflecting torque develops on the coil one is D_1 and the power passes through that element is P_1. Similarly, the torque develops on the coil 2 is D_2 and the power passes through the coil is P_2.

- Deflecting torque of Element $1 \propto P_1$

- Deflecting torque of Element $2 \propto P_2$

The total torque develops in the coil is expressed as:

- Total Deflecting Torque $\propto \left(P_1 + P_2\right) \propto P$

Connections of Three-Phase Wattmeter

Consider the circuit has two Wattmeters. The current coil of both the Wattmeter connects across any two phases say R and Y. The pressure coil of both the Wattmeters connects across the third phase say B.

The mutual interference between the elements of the Three-Phase Wattmeter will affect their accuracy. The mutual interference is the phenomena in which the field of two elements interacts each other. In three-phase Wattmeter, the laminated iron shield is placed between the elements. The iron shield reduces the mutual effect of the element.

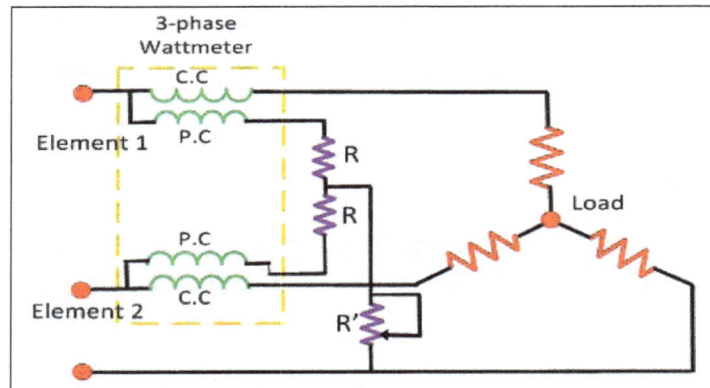

Compensation of Mutual Effects Between two Elements of a Three phase Wattmeter.

The mutual effect can compensate by using the Weston method. In Weston method, the adjustable resistors are used. This resistor compensates the mutual interference that occurs between the elements of three phase wattmeter.

Power Factor Meter

The power factor meter measures the power factor of a transmission system. The power factor is the cosine of the angle between the voltage and current. The power factor meter determines the types of load using on the line, and it also calculates the losses occur on it.

The power factor of the transmission line is measured by dividing the product of voltage and current with the power. And the value of voltage current and power is easily determined by the voltmeter, ammeter and wattmeter respectively. This method gives high accuracy, but it takes time.

The power factor of the transmission line is continuously changed with time. Hence it is essential to take the quick reading. The power factor meter takes a direct reading, but it is less accurate. The reading obtained from the power factor meter is sufficient for many purposes to expect precision testing.

The power factor meter has the moving system called pointer which is in equilibrium with the two opposing forces. Thus, the pointer of the power factor meter remains at the same position which is occupied by it at the time of disconnection.

Types of Power Factor Meter

The power factor meter is of two types. They are:

- Electrodynamometer
  - Single Phase Electrodynammmeter
  - Three Phases Electrodynamometer
- Moving Iron Type Meter
  - Rotating Iron Magnetic Field
  - Number of Alternating Field

Single Phase Electrodynamometer Power Factor Meter

The construction of the single phase electrodynamometer is shown in the figure below. The meter has fixed coil which acts as a current coil. These coils is split into two parts and carry the current under test. The magnetic field of the coil is directly proportional to the current flow through the coil.

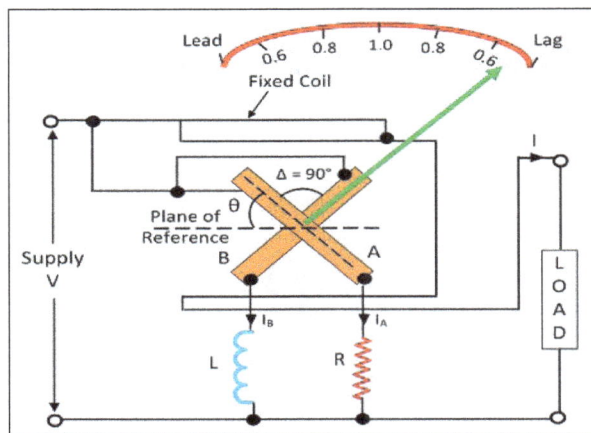

Single Phase Electrodynamometer Type Power Factor Meter.

The meter has two identical pressure coils A and B. Both the coils are pivoted on the spindle. The pressure coil A has no inductive resistance connected in series with the circuit, and the coil B has highly inductive coil connected in series with the circuit.

The current in the coil A is in phase with the circuit while the current in the coil B lag by the voltage nearly equal to 90°. The connection of the moving coil is made through silver or gold ligaments which minimize the controlling torque of the moving system.

The meter has two deflecting torque one acting on the coil A, and the other is on coil B. The windings are so arranged that they are opposite in directions. The pointer is in equilibrium when the torques are equal.

Deflecting torque acting on the coil A is given as:

$$T_A = KVIM \cos\phi \sin\theta$$

Where,

- θ – angular deflection from the plane of reference.

- M_{max} – maximum value of mutual inductance between the coils.

The deflecting torque acting on coil B is expressed as:

$$I_B = KVIM_{max} Cos(90° - \phi) Sin(90° + \phi)$$
$$I_B = KVIM_{max} \cos\phi \sin\theta$$

The deflecting torque is acting on the clockwise direction.

The value of maximum mutual inductance is same between both the deflecting equations.

$$T_A = T_B$$
$$KVIM \cos\phi \sin\theta = KVIM_{max} \cos\phi \sin\theta$$

This torque acts on anti-clockwise direction. The above equation shows that the deflecting torque is equal to the phase angle of the circuit.

Three Phase Electrodynometer Power Factor Meter

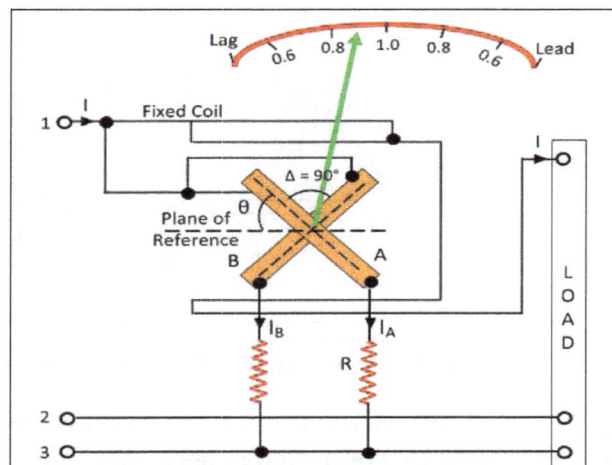

Three Phase Dynamo Type Factor Meter.

The construction of the three phase meter is shown in the figure below. The electrodynamometer is only useful for the balanced load. The moving coil is placed at an angle of 120°. They are connected across different phases of the supply circuit. Both the coil has a series resistance.

The voltage across the coil A is V_{12} and the current across it I_{A1}. The circuit of the coil is resistive, and hence the current and voltage are in phase with each other. Similarly, the voltage V_{13} and the current I_{B1} is in phase with each other.

The phasor diagram of the three phase electrodynamic meter is shown in the figure.

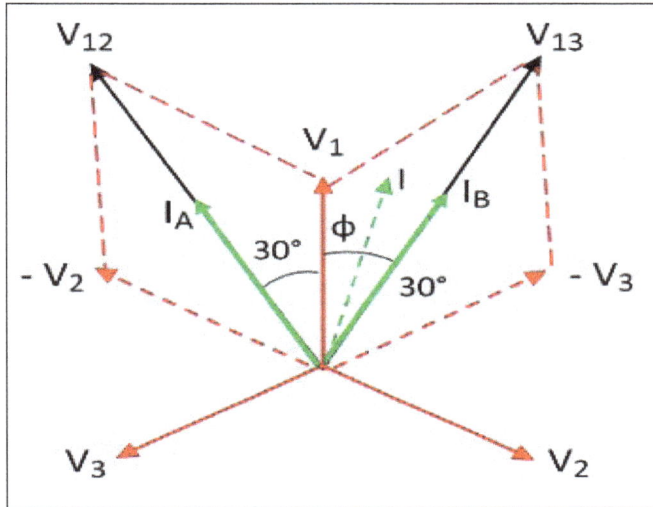

Phasor Diagram of Three Phase Electrodynameter Type Power Factor Meter.

Let,

- Φ – phase angle of the circuit.

- θ – angular deflection from the plane of reference.

Torque acting on coil A is:

$$T_A = KVI_{12}M_{max}Cos(30° + \phi)Sin(60° + \phi)$$
$$T_A = \sqrt{3}KVI_{12}M_{max}Cos(30° + \phi)Sin(60° + \phi)$$

Torque acting on coil B is:

$$T_B = KVI_{12}M_{max}Cos(30° - \phi)Sin(120° + \phi)$$
$$T_B = KVI_{12}M_{max}Cos(30° - \phi)Sin(120° + \phi)$$

The torque T_A and T_B are acting on the opposite directions.

$$Cos(30° - \phi)Sin(120° + \phi) = Cos(30° - \phi)Sin(120° + \phi)$$

Thus the angular deflection of the coil is directly proportional to the phase angle of the circuit.

Moving Iron Power Factor Meter

The moving iron instrument is divided into two categories. They are the rotating magnetic field to some alternating fields.

- Rotating Field Power factor Meter: The following are the essential feature of the rotating magnetic field. The power factor meter has three fixed coils, and their axes are 120° displaced from each other. The axes are intersecting each other. The coils are connected to the three phase supply with the help of the current transformer.

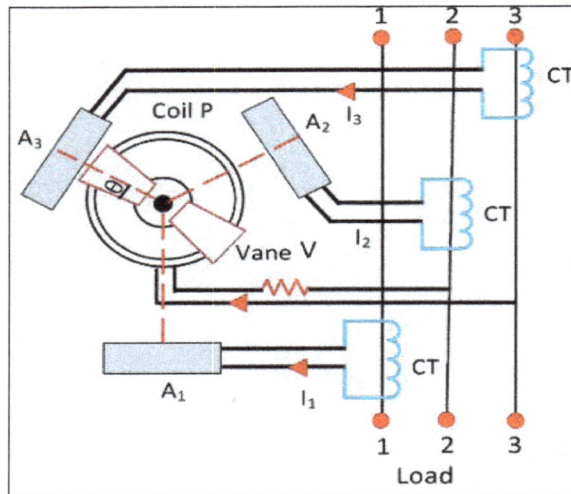

Rotation Field Moving iron Power Factor Meter.

The P is the fixed coil connected in series with the high resistance circuit across the phases 2 and 3. There is an iron cylinder across coil P. The two iron vanes are fixed to the cylinder. The spindles also carry damping vanes and pointer.

The phasor diagram of the power factor meter is shown in the figure.

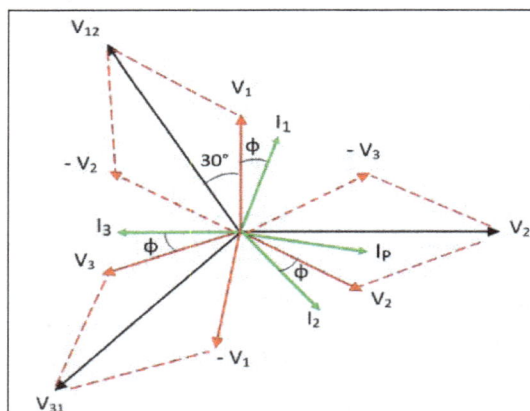

Phasor Diagram for Moving Iron Type Power Factor Meter.

The total torque of the meter is zero for steady state deflection.

$$\left[Cos\left(90°-\phi\right)Sin\left(90°+\phi\right)\right]+Cos\left(330°-\phi\right)Sin\left(210°+\phi\right)+Cos\left(210°-\phi\right)Sin\left(330°+\phi\right)=0$$

The coil P and the iron cylinders generate the alternating flux which interacts with the flux of the fixed coils. The interaction of the coil generates the moving system which determined the phase angle of the current. The vanes of the power factor meter are magnetized by the current of the moving coil which is in phase with the system line voltage.

Advantages of Moving Iron power Power Factor

1. The meter requires large working force as compared to the electrodynamometer type meter.

2. The coils of the moving iron instruments are fixed permanently.

3. The range of the scale extends up to 360°.

4. The construction of the meter is robust and simple.

5. The moving iron instrument is cheap as compared to electrodynamic meter.

Disadvantages of Moving Iron Instrument

1. The loss occurs in the iron part of the meter. The losses depend on the load and the frequency of the meter.

2. The meter has low accuracy.

3. The calibration of the meter is affected because of the variation in supply frequencies, voltage and waveforms etc.

The power factor meter is used for measuring the power factor of the balanced load.

Frequency Meter

Frequency Meter is an instrument for measuring the frequency of periodic processes (oscillations). The frequency of mechanical vibrations is usually measured by means of mechanical vibration frequency meters and by electrical meters equipped with transducers to convert the mechanical vibrations into electrical oscillations.

The simplest mechanical vibration frequency meter operates on the resonance principle and consistsof a series of flexible reeds fastened at one end to a common base. The lengths and masses of the reeds are chosen in such a way that their natural vibration frequencies form a specified discrete scale, from which the value of the frequency being measured is determined. When mechanical vibrations act on the base of the meter, they cause the flexible reeds to vibrate; the highest vibration amplitude is observed on the reed whose natural vibration frequency is equal or close to the value of the frequency being measured.

The frequency of electrical oscillations is measured by means of electromechanical, electrodynamic, electronic, moving-iron, and moving coil frequency meters. The simplest electromechanical type consists of an electromagnet and a series of flexible reeds (as in the mechanical frequency meter) on a common base that is attached to the armature of the electromagnet. The electrical oscillations being measured are fed to the winding of the electromagnet; the armature vibrations thereby produced are transmitted to the reeds, and the value of the frequency being measured is determined from the vibrations.

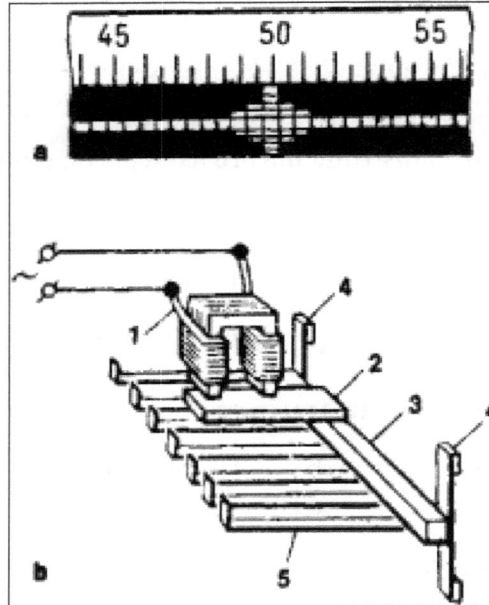

Electromechanical vibration frequency meter: (a) scale, registering a reading of 50 Hz, (b) diagram of the instrument; (1) electromagnet winding, (2) electromagnet armature, (3) base of the frequency meter, (4) elastic supports, (5) reeds.

The principal element in electrodynamic frequency meters is a ratio meter with an oscillatory circuit in one of its branches that is permanently tuned to the average frequency for the measurement range of the given instrument. When connected to an AC circuit, the moving part of the ratio meter is deflected by an angle proportional to the phase shift between the currents in the windings of the ratio meter, which depends on the ratio of the frequency being measured to the resonance frequency of the oscillatory circuit. The measurement error of electrodynamic frequency meters ranges from 10^{-1} to 5×10^{-2}.

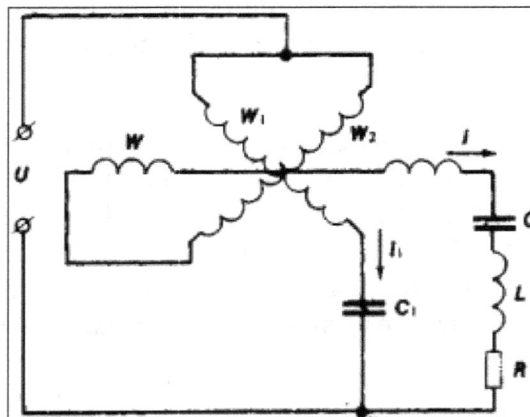

In figure, schematic diagram of an electrodynamic frequency meter: (W) fixed coil of a ratio meter consisting of two identical parts, designed to create a uniform magnetic field; (W_1) and (W_2) moving coils rigidly secured together at an angle of 90°, which interact with coil W; (C), (L), and (R) electrical capacitance, inductance, and resistance, respectively, of the oscillatory circuit; (C_1) capacitor to produce a phase shift of 90° between U and l_1; (U) voltage whose frequency is being measured; (l) and (l_1) currents in the branches of the ratio meter.

The frequency of electromagnetic oscillations in the radio-frequency and microwave frequency ranges is measured by means of electronic frequency meters (wavemeters), such as the resonant, heterodyne, and digital types.

The operation of a resonant type frequency meter is based on the comparison of the frequency being measured with the frequency of natural oscillations in an electrical circuit (or a microwave resonator) that is tuned to resonance with the frequency being measured. The meter consists of an oscillatory circuit with a coupling loop that picks up the electromagnetic oscillations (radio waves), a detector, an amplifier, and a resonance indicator. During measurement, the circuit is tuned by means of a calibrated capacitor (or the plunger of a microwave resonator) to the frequency of the electromagnetic oscillations being picked up until resonance is achieved, as shown by the greatest deflection of the pointer on the indicator. The measurement error ranges from 5 × 10^{-3} to 5 × 10^{-4}.

In heterodyne frequency meters, the frequency being measured is compared with a known frequency (or one of its harmonics) produced by an oscillator, or heterodyne. As the heterodyne frequency is tuned to the frequency of the oscillations being measured, beats occur at the output of a mixer (in which the frequencies are compared); after amplification the beats are indicated by the pointer on an instrument, by an earphone, or sometimes by an oscilloscope. The relative error of heterodyne frequency meters ranges from 5 × 10^{-4} to 5 × 10^{-6}.

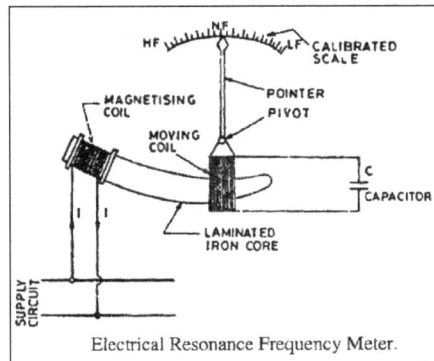

Electrical Resonance Frequency Meter.

Schematic diagram of a resonant-type frequency meter

Digital frequency meters (frequency counters) are now widely used. Their operation involves a count of the number of periods in the oscillations being measured during a specified time interval. Frequency counters consist of a device that converts the sinusoidal voltage of the frequency being measured into a train of unidirectional pulses, a gate for the pulses that opens for a certain time interval (usually from 10^4 to 10 sec), an electronic counter that registers the number of pulses at the gate output, and a digital display. Modern digital frequency counters operate over a frequency range from 10^4 to 10^9 hertz with a relative measurement error from 10^{-9} to 10^{-11} and a sensitivity of 10^{-2} volt. Such devices are used primarily for testing radio equipment and, with various measuring transducers, for measuring temperature, vibrations, pressure, strain, and other physical quantities.

Primary and secondary frequency standards, which have an error in the range from 10^{-12} to 5×10^{-14}, function as a type of high accuracy reference frequency meters. The rotational speed of shafts in machines and mechanisms is measured with a tachometer.

Digital Frequency Meter

Digital frequency meter is a general purpose instrument that displays the frequency of a periodic electrical signal to an accuracy of three decimal places. It counts the number events occurring within the oscillations during a given interval of time. As the present period gets completed, the value in the counter display on the screen and the counter reset to zero. Various types of instruments are available which operates at a fixed or variable frequency. But if we operate any frequency meter at different frequencies than the specified range, it could carry out abnormally. For measuring low frequencies, we usually use deflection type meters. The deflection of the pointer on the scale shows the change in frequency. The deflection type instruments are of two types: one is electrically resonant circuits, and other is ratio meter.

Operating Principle of Digital Frequency Meter

A frequency meter has a small device which converts the sinusoidal voltage of the frequency into a train of unidirectional pulses. The frequency of input signal is the displayed count, averaged over a suitable counting interval out of 0.1, 1.0, or 10 seconds. These three intervals repeat themselves sequentially. As the ring counting units reset, these pulses pass through the time-base-gate and then entered into the main gate, which opens for a certain interval. The time base gate prevents a divider pulse from opening the main gate during the display time interval. The main gate acts as a switch when the gate is open; pulses are allowed to pass. When the gate is closed, pulses are not allowed to pass that means the flow of pulses get obstructed.

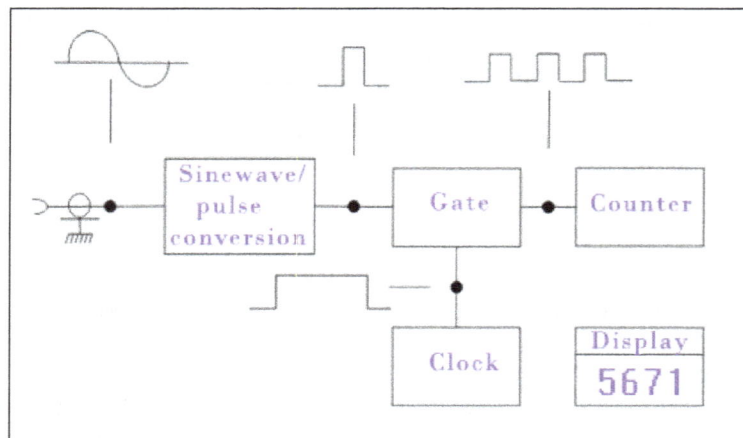

The functioning of the gate is operated by the main-gate flip-flop. An electronic counter at the gate output that counts the number of pulses passed through the gate while it was open. As the main gate flip-flop receives next divider pulse, the counting interval ends, and divider pulses are locked out. The resultant value displayed on a display screen which has the ring counting units of scale-of-ten circuits and each unit couples to a numeric indicator, which provides the digital display. As the reset pulse generator is triggered, ring counters get reset automatically, and the same procedure starts again.

The range of modern digital frequency meter is between the range from10^4 to 10^9 hertz. The possibility of relative measurement error ranges between from 10^{-9} to 10^{-11} hertz and a sensitivity of 10^{-2} volt.

Use of Digital Frequency Meter

- For testing radio equipment;

- Measuring the temperature, pressure, and other physical values;

- Measuring vibration, strain;

- Measuring transducers.

Flux Meter

The meter which is used for measuring the flux of the permanent magnet such type of meter is known as the flux meter. The fluxmeter is the advanced form of the ballistic galvanometer which has certain advantages like the meter has low controlling torque and heavy electromagnetic damping.

Construction of Flux Meter

The construction of the fluxmeter is shown in the figure below. The fluxmeter has a coil which is freely suspended by the help of the spring and the single silk thread. The coil moves freely between the poles of the permanent magnet.

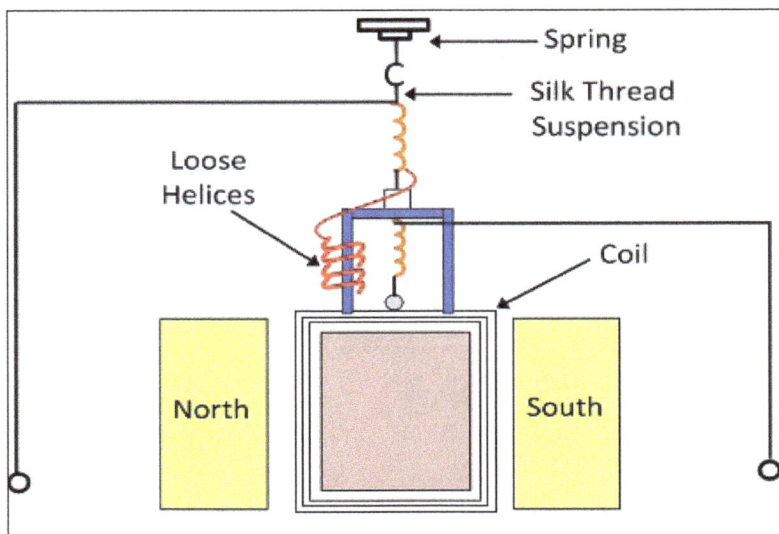

Flux Meter.

The current enters into the coil with the help of the helices which is very thin and made from the annealed silver strips. This current reduces the controlling torque to the minimum value. The air friction damping of the coil is negligible.

Operation of Flux Meter

The terminals of the fluxmeter are connected across the search coil as shown in the figure below. The flux linking with the coil is varied by either removing it from the magnetic field or by reversing the field of the magnet. The change of the flux induces the electromotive force in the coil. This emf induces the current in the search coil and send it through the flux meter. Because of the current, the pointer of the fluxmeter deflects, and their deflection is directly proportional to the change in the value of flux linkages.

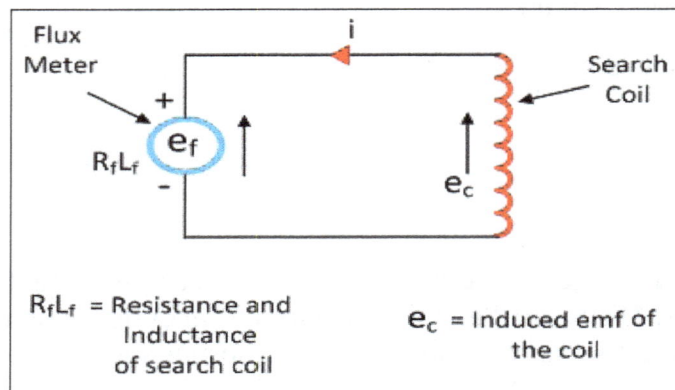

Flux Meter with Search coil.

As, the variation of the flux linkages reduces, coil stop moving because of their high electromagnetic damping. The high electromagnetic damping is because of the low resistance circuit between the fluxmeter and the search coil.

Advantages of Fluxmeter

The following are the advantages of fluxmeter:

- The fluxmeter is portable.

- The scale of the fluxmeter is calibrated in Weber meters.

- The deflection of the coil is free from the time taken by the flux to change.

Disadvantages

The only disadvantage of the fluxmeter is that it is less sensitive and accurate as compared to the flux meter.

References

- Moving-coil-galvanometer, moving-charges-and-magnetism, physics: toppr.com, Retrieved 2 February, 2019

- Galvanometer: circuitglobe.com, Retrieved 22 July, 2019

- Moving-coil-galvanometer, moving-charges-and-magnetism, physics: toppr.com, Retrieved 29 March, 2019

- Working-principle-of-voltmeter-and-types-of-voltmeter: electrical4u.com, Retrieved 23 January, 2019

- Multimeter-types-and-applications: elprocus.com, Retrieved 21 June, 2019

- Energy-meter: circuitglobe.com, Retrieved 25 April, 2019

- What-is-a-wattmeter: wisegeek.com, Retrieved 5 March, 2019

- How-does-a-wattmeter-work-back-to-basics: electrical-equipment.org, Retrieved 15 June, 2019

- Three-phase-wattmeter: circuitglobe.com, Retrieved 13 May, 2019

- Frequency-Meter: thefreedictionary.com, Retrieved 3 February, 2019

- Digital-frequency-meter: electrical4u.com, Retrieved 11 August, 2019

Permissions

All chapters in this book are published with permission under the Creative Commons Attribution Share Alike License or equivalent. Every chapter published in this book has been scrutinized by our experts. Their significance has been extensively debated. The topics covered herein carry significant information for a comprehensive understanding. They may even be implemented as practical applications or may be referred to as a beginning point for further studies.

We would like to thank the editorial team for lending their expertise to make the book truly unique. They have played a crucial role in the development of this book. Without their invaluable contributions this book wouldn't have been possible. They have made vital efforts to compile up to date information on the varied aspects of this subject to make this book a valuable addition to the collection of many professionals and students.

This book was conceptualized with the vision of imparting up-to-date and integrated information in this field. To ensure the same, a matchless editorial board was set up. Every individual on the board went through rigorous rounds of assessment to prove their worth. After which they invested a large part of their time researching and compiling the most relevant data for our readers.

The editorial board has been involved in producing this book since its inception. They have spent rigorous hours researching and exploring the diverse topics which have resulted in the successful publishing of this book. They have passed on their knowledge of decades through this book. To expedite this challenging task, the publisher supported the team at every step. A small team of assistant editors was also appointed to further simplify the editing procedure and attain best results for the readers.

Apart from the editorial board, the designing team has also invested a significant amount of their time in understanding the subject and creating the most relevant covers. They scrutinized every image to scout for the most suitable representation of the subject and create an appropriate cover for the book.

The publishing team has been an ardent support to the editorial, designing and production team. Their endless efforts to recruit the best for this project, has resulted in the accomplishment of this book. They are a veteran in the field of academics and their pool of knowledge is as vast as their experience in printing. Their expertise and guidance has proved useful at every step. Their uncompromising quality standards have made this book an exceptional effort. Their encouragement from time to time has been an inspiration for everyone.

The publisher and the editorial board hope that this book will prove to be a valuable piece of knowledge for students, practitioners and scholars across the globe.

Index